INTRODUCTION TO GC-MS COUPLING

STÉPHANE BOUCHONNET

INTRODUCTION TO GC-MS COUPLING

CRC Press
Taylor & Francis Group
Boca Raton London New York

CRC Press is an imprint of the
Taylor & Francis Group, an **informa** business

This is a translation from the French book *La SPECTROMÉTRIE de MASSE; en COUPLAGE avec la CHRO-MATOGRAPHIE en PHASE GAZEUSE.*

CRC Press
Taylor & Francis Group
6000 Broken Sound Parkway NW, Suite 300
Boca Raton, FL 33487-2742

© 2013 by Taylor & Francis Group, LLC
CRC Press is an imprint of Taylor & Francis Group, an Informa business

No claim to original U.S. Government works

Printed on acid-free paper
Version Date: 20130125

International Standard Book Number-13: 978-1-4665-7251-5 (Paperback)

Library of Congress Cataloging-in-Publication Data

Bouchonnet, Stéphane.
    Introduction to GC-MS spectrometry / Stéphane Bouchonnet.
        p. cm.
    Includes bibliographical references and index.
    ISBN 978-1-4665-7251-5 (pbk.)
    1. Gas chromatography. 2. Mass spectrometry. I. Title. II. Title: Introduction to gas chromatography-mass spectrometry.

QD79.C45B68 2013
543'.65--dc23   2012037919

Visit the Taylor & Francis Web site at
http://www.taylorandfrancis.com

and the CRC Press Web site at
http://www.crcpress.com

To my family, which is patiently organized around a daddy who spends a lot of time writing books and drawing hives.[*]

---

[*] Aromatic rings as seen by children.

Contents

Introduction

Most of the analytical chemistry laboratories are equipped with a gas chromatograph coupled with a mass spectrometer today. This technique is generally referred to as GC-MS (gas chromatography–mass spectrometry). The production of mass spectrometer manufacturers has undergone a huge growth during the past 20 years and sales of GC-MS apparatus are still expanding worldwide. GC-MS finds applications in fields as diverse as the food processing industry, medicine, pharmacology, and environmental analysis.

This handbook addresses every user of GC-MS coupling, novice or experienced. The idea of its writing arose from a report that few works were dedicated to mass spectrometry used as a detector for gas chromatography. Besides the criticism of these books for their generally high complexity, most present a very theoretical approach to mass spectrometry that is difficult to understand and disheartening for new users. The practical aspects such as the development of a dosage method or the use of mass spectra databases are almost systematically neglected in books devoted to mass spectrometry although these are the issues that novice users ponder daily.

The ambition of this book is to emphasize the practical aspects of the technique without neglecting the fundamental theoretical aspects. The majority of quadrupole users have never studied the Mathieu equations. Fortunately, this does not prevent most of them from using their mass spectrometers effectively. Here, the purpose is not so much to calculate the trajectory of an ion in a field but to learn to choose the most successful device for a given application and understand why it is necessary to avoid quantifying molecules by external calibration.

The framework of this book is a series of mass spectrometry courses conducted for more than 12 years for private industry and at the University of Paris XI (France). Theoretical demonstrations are supported by numerous examples because experience has shown the utility of examples in aiding understanding of the material. This book also discusses the problems and "traps" usually met by novice users and tries to anticipate the most common mistakes and to bring solutions to them.

This book includes nine chapters. The first chapter presents the main principles of gas chromatography and its specifics in use with MS detectors. Gas chromatography is a physico-chemical technique of separation that is widely used. It has been studied in universities and engineering schools for decades and is generally well known by chemists and especially by analytical chemistry specialists. This technique and its multiple applications have been the subjects of many books. If the equipment has been the object of some technical improvements during the last decade, the technique has not significantly evolved as quickly. Nevertheless, two innovative concepts have recently emerged: fast chromatography and multidimensional chromatography. These techniques will undoubtedly be used more and more in the future but are not routinely used at present; they are briefly described in this chapter.

This book deals first with mass spectrometry; therefore, the chapter devoted to gas chromatography presents scholarly reminders but is not expected to be exhaustive.

It aims to reiterate some fundamental concepts necessary to understand how a chromatographic separation is carried out. At the risk of seeming overly popularized to experts in the field, this chapter aims to be as clear and simple as possible, without equations or theoretical considerations. The practical aspects are emphasized along with the specifics of gas chromatography when it is coupled with mass spectrometry.

The second chapter is dedicated to some generalities on the technique of mass spectrometry. In particular, it provides an insight into the vacuum necessary to ensure efficient operation of mass spectrometers. The various types of MS detectors that can be set onto a GC-MS device are also presented in this chapter.

The third chapter is dedicated to the ionization modes that can be operated in GC-MS: electron ionization as well as positive and negative chemical ionization. We explain the principles of operation of internal and external sources and compare the performances achieved by each ionization mode.

The fourth chapter presents the various types of analyzers used in GC-MS and their functioning principles. The main part of the chapter is devoted to quadrupolar analyzers since quadrupolar filters and ion traps equip the great majority of GC-MS devices.

Chapter 5 is dedicated to the various modes of acquisition: selected ion monitoring (SIM), tandem mass spectrometry (MS/MS), and others. Their implementation, advantages, and limits are detailed.

The sixth chapter presents a comparison of the performances and the limitations of the quadrupolar analyzers since experience has shown that the choice between these devices is often very difficult for novice users. The chapter is not intended to promote a given type of system but seeks to discard some generally accepted ideas and offer the most complete and objective panorama of analyzers available on the market.

Chapter 7 is dedicated to quantification using GC-MS coupling. The techniques of dosage are generally often ignored in the books devoted to mass spectrometry so that quantification by GC-MS remains a puzzle for numerous chemists. Laboratory audits show that it is when attempting to perform quantitative analysis that operators have to face the main difficulties and often make mistakes. An example of the development of a quantification method is presented.

The eighth chapter deals with the use of databases in mass spectrometry. We compare the performances and limitations of laboratory-made and commercial mass spectra databases; a database research algorithm is also described. The objective of this chapter is to encourage a critical approach to databases for a relevant and objective use.

Finally, the ninth chapter is dedicated to mass spectra interpretation. Concepts of chemistry such as inductive and mesomeric effects required to understand dissociation pathways are covered. The main fragmentation mechanisms, simple cleavages, and rearrangements are presented along with isotopic distributions, the nitrogen rule, and a global strategy for mass spectra interpretation.

The principles of GC-MS are perfectly understood today. Conversely, high performance liquid chromatography coupled with mass spectrometry (LC-MS) using the electrospray process remains the object of academic research. Nevertheless, developing dosage methods by GC-MS remains a real analytical challenge. In particular, sample preparation constitutes a fundamental stage of the analysis. Sample preparation is often a complex process performed over numerous stages: extraction of analytes from the matrix, purification, chemical derivation, concentration, centrifugation, etc. There are so many techniques of sample preparation (and so many variants in each technique) that each of them deserves a chapter. Sample preparation is therefore dealt with briefly in this work, mainly through examples. That is why analysts who wish to acquire skills in sample preparation should refer to specialized works in this domain.

This work was conceived as a handbook to enable readers to find as many answers as possible to their questions without having to always consult other books or articles even if increasing knowledge through additional reading is obviously a good thing. Anyone who has already spent hours reading numerous articles and following footnotes from one article to the next knows how boring and disheartening this experience is and how incompatible it is with the time allowed for bibliographical research in most analytical laboratories!

This work cites bibliographical references, especially when a technical explanation cannot claim to be exhaustive (the case of analyzers using magnetic sectors, for example) or when a theoretical development concerns only a marginal number of users (multidimensional GC, for example). There are not many bibliographical references in this book, mainly for two reasons. First, a multitude of scientific articles dedicated to mass spectrometry have been published during recent years. Thus, it is impossible to read the thousands of publications on the subject and achieve an objective and rigorous selection among them. The second reason is that numerous scientific articles are accessible on the web and are easy to select with current search engines.

This book is necessarily imperfect. The progress of the technique is so rapid that certain technical aspects may seem obsolete only a few months after publication. Any comments and suggestions to improve this book are welcome at the email address of the author: stephane.bouchonnet@dcmr.polytechnique.fr

Abbreviations

amu	Atomic mass unit
ASE	Accelerated solvent extraction
BSA	N,O-bis(trimethylsilyl)acetamide
BSTFA	N,O-bis(trimethylsilyl)trifluoroacetamide
CAS	Chemical abstract service
CI	Chemical ionization
CV	Coefficient of variation
ECD	Electron capture detector
EA	Electron affinity
EI	Electron ionization
FID	Flame ionization detector
GC	Gas chromatography
GC-MS	Gas chromatography coupled with mass spectrometry
HPLC	High performance liquid chromatography
I	Inductive effect: +I, electron releasing; –I, electron withdrawing
ICR	Ionic cyclotronic resonance
IE	Ionization energy
IP	Ionization potential
LC	Liquid chromatography (see HPLC)
LC-MS	Liquid chromatography coupled with mass spectrometry
LOD	Limit of detection
LOQ	Limit of quantification
LSD	Lysergic diethylamide acid
M	Mesomeric effect: +M, positive mesomeric effect; –M, negative mesomeric effect
MCP	Microchannel plate
McL	McLafferty rearrangement
MRM	Multiple reaction monitoring
MS	Mass spectrometry
MS/MS	Tandem mass spectrometry
MSTFA	n-methyl-n-(trimethylsilyl)trifluoroacetamide
NPD	Nitrogen phosphorus detector
MW	Molecular weight
PA	Proton affinity
PBDE	Polybromodiphenylether
PCB	Polychlorobenzene
PPB	Part per billion

PSI	Pound per square inch
RDA	Retro Diels-Alder reaction
NMR	Nuclear magnetic resonance
SIM	Selected ion monitoring
SIR	Selected ion recording
SIS	Selected ion storage
SPE	Solid phase extraction
SPME	Solid phase micro extraction
SRM	Selected reaction monitoring
TMCS	Trimethylchlorosilane
TMSDEA	n-trimethylsilyldiethylamine
TMSI	1-(trimethylsilyl)imidazole
TOF	Time-of-flight
TQ	Triple quadrupole
TS	Transition state

Acknowledgments

With warm thanks to Sonia Ashbee and Kim Halle who assisted me with the translation from French.

With huge thanks to Yannick Hoppilliard who was my thesis director 20 years ago and inspired me to learn and teach mass spectrometry; to Guy Bouchoux who was my university professor for mass spectrometry at Paris XI; to Henry Edouard Audier who provided me the opportunity to join the Laboratoire des Mécanismes Réactionnels at the Ecole Polytechnique, where I still work today and who always provided me very helpful advice; and to Gilles Ohanessian, currently director of this research laboratory, for his encouragement and support.

Thanks also to Tom Norton, Terry McMahon, Ivan Ricordel, Christian Staub, and Delphine Brulé for their kindness during our fruitful scientific exchanges.

I am also obliged to all my students, far too numerous to be named here, who made these last years pass by too fast.

1 Concepts of Gas Chromatography

1.1 GENERAL POINTS

1.1.1 GENERAL PRINCIPLE

The role of a chromatograph is to separate the compounds of a mixture. When speaking of chromatography in the gas phase (or GC for gas chromatography), molecules that are to be separated volatilize and mix with a gas. This gas, called *carrier gas*, constitutes the mobile phase; it transports the analytes into an analytical column whose inner surface is covered with a chemical film (or stationary phase). The molecules are separated in time because they migrate in the column at various speeds; the length of each analyte course depends on its volatility and the interactions between the molecule and the chemical film.

The analytes are detected at their exit from the column. Each molecule is characterized by a retention time that corresponds to the time that passed between the injection of the analyte and its arrival at the detector. Concerning volatility, the separation principle is simple: the speed at which a compound migrates in the column relates to its boiling point. The boiling point is a thermodynamic measure that depends mainly on two factors: the molecular weight and polarity. A compound is as volatile as its molecular weight and polarity are low. The interaction between analytes and the stationary phase are more complex; one must chose the type of chemical film to use according to the nature of the molecules that need to be separated.

The principle of gas chromatography can be approached in the following manner. Let's imagine that various pieces of wood (symbolizing the analytes) are dropped from a bridge into a river (the analytical column). Water symbolizes the carrier gas. According to their weight, the pieces of wood will float more or less. Depending on their size, shape, and roughness, they will hit the river bank (the chemical film that constitutes the stationary phase) at various times and some will attach to it. Thus, the different pieces of wood will reach a bridge downstream at different times.

The river's descent speed depends not on the properties of each piece of wood, but on the size of the river, the speed of the current, the composition of the river bank, and the distance between the bridges. The pieces of wood that are too heavy to float will not reach the second bridge nor will those whose shapes make them catch onto the river bank and contribute to the fouling of the river. It is therefore advisable, before initiating an analysis by gas chromatography, to impose the sample to a preparation aiming to remove a maximum of molecules that are not sufficiently volatile. For a theoretical approach to gas chromatography, one can refer to the work of Grob and Barry titled *Modern Practice of Gas Chromatography*.[1]

1.1.2 SUITABLE MOLECULES FOR GC ANALYSIS

Gas chromatography is confined to the analysis of relatively volatile and thermally stable compounds. It is difficult to define exactly its limits in terms of molecular weights but cases of analysis of compounds having molecular weights approaching 1000 amu are exceptional. However, analyses of decabromodiphenylether (formula $C_{12}Br_{10}O$, average molecular weight equal to 960 amu), a flame retardant frequently used in the plastic and textile industries, have been reported.[2] The most common use of gas chromatography therefore essentially concerns small to average size and low polarity molecules.

Among the main sectors that use GC-MS coupling, we can mention environmental analysis, the perfume, cosmetic, and aroma industries, pharmaceutics, and toxicology. With rare exceptions, all GC-analyzable molecules are also analyzable by GC-MS because the mass spectrometer sources used in GC-MS do not impose a particular limit on the natures of analytes.

1.1.3 GC-MS VERSUS LC-MS

The place of GC-MS coupling within analytical techniques has changed over the last few years with the advent of instruments coupling liquid chromatography with mass spectrometry (LC-MS or liquid chromatography–mass spectrometry).[3] In some cases, the choice of a definite technique is not difficult: GC-MS is suitable for very volatile molecules (polycyclic aromatic hydrocarbons, dioxins, furans, residual solvents) and LC-MS for large size molecules (peptides, proteins, industrial polymers) and very polar molecules (hydrophilic medicines, salts). In other cases, the choice between GC-MS and LC-MS is more complex. These cases are numerous; they concern all average-sized molecules (molecular weights between 150 and 600 amu, approximately) such as pesticides, estrogens, most psychotropic drugs and narcotics analyzed in toxicology laboratories. Many studies conducted on pesticides illustrate the complementarity of these two techniques.[4–6]

A brief comparison of the GC-MS and LC-MS techniques highlights the following advantages and inconveniences. GC-MS is less costly; methodological procedures are generally simple even for the analysis of complex mixtures; and the adaptation of traditional GC methods for GC-MS is easy. By contrast, using this technique imposes frequently to derive chemicals (refer to next paragraph) that are so polar that they are poorly volatile; it complicates the sample preparation. LC-MS analysis generally avoids the need for chemical derivation of the sample; it is a shorter procedure than GC-MS analysis and often supplies inferior detection limits insofar as it allows the analysis of bigger sample volumes. By contrast, the methodological development is often more complex and longer in LC-MS than GC-MS. It is the same for validation because LC-MS is frequently subject to spectral suppression phenomena that considerably complicate analysis in complex matrices.[7]

1.1.4 CHEMICAL DERIVATION

Chemical derivation is frequently used to analyze by gas chromatography molecules of low volatility because they are too polar. Exchangeable hydrogen atoms (in particular

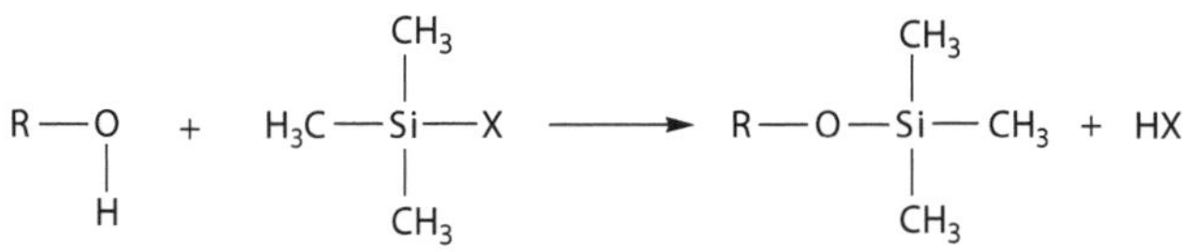

FIGURE 1.1 Silylation of an alcohol with BSTFA.

the hydrogens of carboxylic, amino, and alcohol functions) are responsible for strong interactions such as hydrogen binding with the glass insert (also called the liner) of the chromatograph's injector and/or with the analytical column's stationary phase. These interactions lead to chromatographic peak tailing that degrades chromatographic resolution and reduces the signal-to-noise ratios for the corresponding peaks.

Chemical derivation can also help to avoid thermal degradation of the thermolabile compounds in the injector or the column.[8,9] Dozens of chemical derivation protocols exist. Below, four of the most used reactions are presented: silylation, methylation, esterification, and acetylation.

1.1.4.1 Silylation

Silylation is without a doubt the most employed derivation procedure today. It is particularly effective with alcohols, phenols, sugars, amines, thiols, steroids, and carboxylic acids.[10] The reaction of an alcohol presented in Figure 1.1 replaces an exchangeable hydrogen with a silicon. The operational conditions depend obviously on the analytes and the reagents used, but most silylation reactions are performed in an oven, at temperatures ranging from 60 to 80°C and with reaction times ranging from 20 to 30 minutes.

Among the main commercial silylation agents, we can mention N,O-bis(trimethylsilyl)trifluoroacetamide (BSTFA) to which is generally added 1% of trimethylchlorosilane (TMCS), the latter playing the role of a catalyst in the reaction, n-methyl-n-(trimethylsilyl)trifluoroacetamide (MSTFA), n-trimethylsilyldiethylamine (TMSDEA), N,O-bis(trimethylsilyl)acetamide (BSA), generally used in mixture with trimethylchlorosilane, and 1-(trimethylsilyl)imidazole (TMSI). These molecules are presented in Figure 1.2.

The silylation agent also often fills the function of a solvent for the derivation reaction. Because it is volatile, it can also be used as an injection solvent for chromatography.

1.1.4.2 Methylation

Methylation is generally used for the alkylation of phenols (Figure 1.3). Iodomethane (CH_3I) is the most used reactive. Dimethylsulfate $(CH_3O\text{-}SO_2\text{-}OCH_3)$ and dimethylcarbonate $(CH_3O\text{-}CO\text{-}OCH_3)$ are also used regularly. Methyltriflate $(CF_3O\text{-}SO_2\text{-}OCH_3)$ and methylfluorophosphonate $(FSO_2\text{-}OCH_3)$ are more reactive but are less used because they are more toxic. The reaction can be catalyzed by Li_2CO_3.[11]

1.1.4.3 Esterification

Very polar carboxylic acids are esterified in esters before their analysis by GC-MS. In the Fisher esterification process, the reactive used is an alcohol, usually methanol or ethanol, which also has the role of solvent in the reaction. This reaction (Figure 1.4)

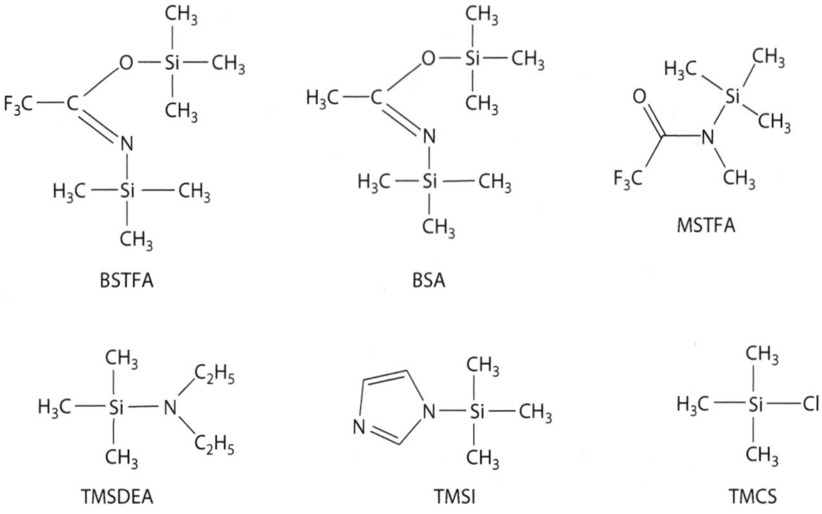

FIGURE 1.2   Silylating agents and generally associated catalysts.

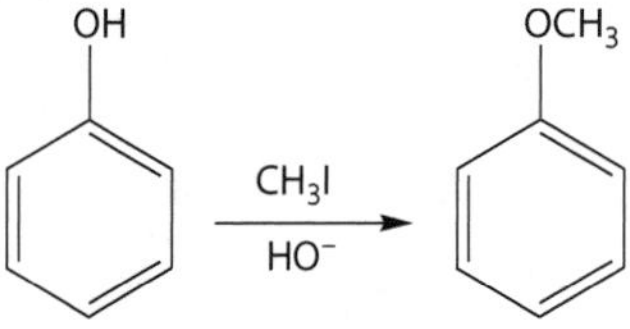

FIGURE 1.3   Methylation of phenol with iodomethane.

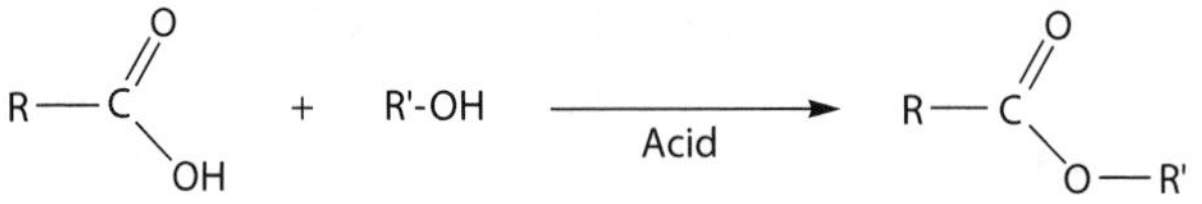

FIGURE 1.4   Esterification of a carboxylic acid.

is catalyzed under acidic conditions, with the addition of sulfuric acid, for example. It can also be performed directly in the injector of the chromatograph by injecting a methanolic solution of carboxylic acids at high temperature.

1.1.4.4   Acetylation

Like silylation, acetylation is effective as a means to substitute exchangeable hydrogens in alcohols, phenols, and primary and secondary amines. The reagents used for acetylation are acid anhydrides such as acetic anhydride, propionic anhydride, or trifluoroacetic anhydride. The reaction temperatures vary between 50 and 150°C and the reaction time from 15 minutes to 2 hours. As for silylation, all traces of water must be avoided. As an example, dopamine acetylation by trifluoroacetic anhydride is presented in Figure 1.5.[10]

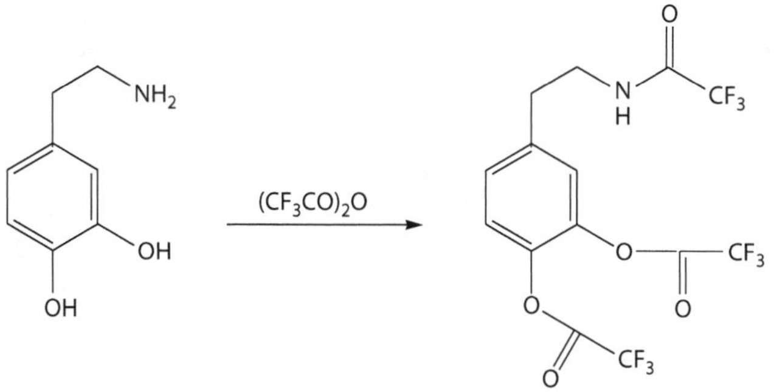

FIGURE 1.5 Acetylation of dopamine.

1.2 INSTRUMENTATION

A gas chromatograph is composed of three elements: an injector, a capillary column situated in an oven, and a detector. Among the several types of detectors including the well-known flame ionization detector (FID), the electron capture detector (ECD), and the specific nitrogen and phosphorus detector (NPD), the mass spectrometer tends to be favored above all because it is the only one to provide structural information on analyzed compounds and allow the quantification of compounds not separated by chromatography.

1.2.1 INJECTOR

1.2.1.1 Carrier Gas

The injector is a heated zone where the sample solution is introduced via a syringe to be then vaporized and mixed with the carrier gas which is the mobile phase. Except for rare applications, the carrier gas used in GC-MS is helium. If nitrogen and hydrogen are often used when a chromatograph is equipped with an FID, ECD, or NPD detector, these gases are rarely used with a mass spectrometer. They are not compatible with the use of an ion trap analyzer with internal ionization where the carrier gas also serves as cooling gas in the mass spectrometer (see Chapter 4). The inert carrier gas has no other function but to allow the elution of the compounds in the analytical column.

1.2.1.2 Gas Flow

The viscosity of a gas varies with the temperature. The carrier gas flow in the analytical column drops with the increase of oven temperature if the pressure at the entrance of the column stays constant during the chromatographic analysis. This results in a decrease in the speed of elution of the analytes, leading to an enlargement of the peaks, degrading the quality of the chromatographic profile.

That is why modern injectors are equipped with electronic flow regulators. These devices adjust the carrier inlet (or column head) pressure according to the temperature of the oven, in order for the gas flow in the column to be constant, which considerably improves the performance of the chromatograph.

The gas flow conditions the pressure within the source of the mass spectrometer. Thus, this pressure results from a balance between the molecules arriving in the source via the chromatograph and the molecules exiting by aspiration (constantly or almost) exerted by the pumping system. The gas flow constraints differ according to the mass spectrometer used. They are low for the quadrupoles and more drastic with ion traps using internal ionization for which the carrier gas plays a fundamental role in ion storage (see Chapter 4). In practice, good results are generally obtained with a flow of about 1 to 2 mL/min for 30 to 60 m long analytical columns with internal diameters of 0.25 mm.

1.2.1.3   Injection in Split Mode

Injection in the split mode is used for the analysis of concentrated solutions. The injection is performed at a high temperature, above the boiling points of the analytes and solvent. The sample is rapidly inserted in the injector where it is instantly vaporized and mixed with the carrier gas. A solenoid valve controls the proportion of the gas mixture (split ratio) that escapes through a leak. This procedure allows one to evacuate of part of the gas flow to decrease the amount of sample that enters the column, thus avoiding the saturation of the stationary phase (refer to Section 1.2.2.1).

Although easy to put into place, the split mode presents an important discrimination problem when the solution to be analyzed contains compounds with very heterogeneous volatilities. The less well vaporized "heavy" products tend to escape through the leak whereas the "light" products mainly remain in the injector.

1.2.1.4   Injection in Splitless Mode

Injection in the splitless mode is used to introduce analytes into a diluted solution. The valve is closed generally for 1 to 2 min after the injection to allow a maximum amount of analyte to enter the column. The valve is then opened to purge the injector of any possible residues. The sample is injected at such a temperature that the solvent and solutes are instantly vaporized in the glass insert.

During the injection, the temperature of the oven is 20 to 30°C below the solvent boiling point in order to condense it and to trap the molecules in the column header. First, the solvent plays the role of stationary phase in relation to the different constituents of the mixture. The solvent polarity must therefore be compatible with that of the stationary phase for the solvent to spread homogeneously in the column header. Because of its high retention power, this condensed phase allows the slowing of the volatile molecules until they are swept by the carrier gas.

This injection mode is the most used because it is efficient and easy to implement. It should be considered, however, that the quantity of sample really introduced into the column usually does not exceed 75 to 80% of the injected quantity. Figure 1.6 illustrates an injector that allows split and splitless injection.

1.2.1.5   On-Column Injection

More efficient than splitless in terms of repeatability and sensitivity, on-column injection means directly introducing the analyte solution into the column or into a

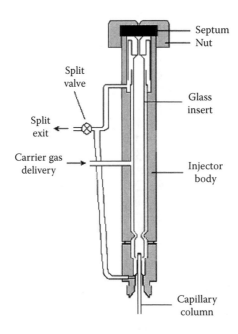

Septum
Nut

Split valve

Glass insert

Split exit

Carrier gas delivery

Injector body

Capillary column

FIGURE 1.6 Representation of an injector for split and splitless injection.

pre-column whose function is to allow the injection of a large volume sample and/or to protect the analytical column from possible matrix pollutants. Pre-column length ranges from 1 to 10 m and its polarity is compatible with the injection solvent or is chemically neutral. Its internal diameter must be sufficient (generally 0.53 mm) for the syringe needle to enter it. Because the solution is injected directly into the pre-column, it is considered that in effect the whole sample is introduced.

This type of injection allows the introduction of much higher sample volumes than those in split or splitless procedures. The volume depends on the length of the pre-column; it can reach several dozens of milliliters, unlike the 0.1 to 2.0 μL generally injected in split and splitless methods.

The injection is mostly done cold: the sample is introduced into the column header or pre-column in its liquid form before being rapidly vaporized. Indeed, in the case of a hot injection, the brutal expansion of the injected solution volume would lead to such a high pressure in the column that the gas would flow back toward the injector. The very efficient on-column injection mode is recommended for trace analysis.

Many users, however, have related problems of leakage in the press-fit connection between the pre-column and the capillary column. The press-fit connection uses a small glass tube to connect two columns whose diameters can be different or identical. Each column is pushed into an extremity of the tube. Sealing is obtained by heating the column. The softening of the polyimide film that composes the outer part of the capillary column (refer to Section 1.2.2.1) leads to its adherence to the outer glass panel. Unfortunately, the sealing of the connection lasts only for a limited time and it is necessary to renew the assembly regularly.

1.2.1.6   Ross Injector

The Ross injector is more scarcely used than those mentioned previously and it allows the injection of large volumes. It is composed of a glass needle approximately 10 cm long placed in a tube equipped with a leak valve and drilled with two lateral orifices (one for the arrival of the carrier gas and the other for the injection of the sample). The sample is placed on the needle. The carrier gas flow allows the evaporation of the solvent that is eliminated by the leak before putting down the needle into the injection port where the sample is vaporized.

This injection mode is particularly useful for trace analysis in environmental matrices. It allows the injection of large sample volumes and successive introductions in order to concentrate analytes on the needle; it also allows avoidance of solvent injection, thus increasing the lifetime of the analytical column. However, it does not allow analysis of very volatile products as they evaporate with the solvent.

1.2.2   Oven and Capillary Column

The oven contains the key element of chromatographic separation: the analytical column. The columns currently used in GC-MS are almost exclusively capillary columns—so named because they are very long and thin. They are both robust and efficient in terms of chromatographic resolution.

1.2.2.1   Composition of a Capillary Column

Figure 1.7 represents a capillary column trunk under sagittal sectioning. The column is composed of a fused silica tube whose internal surface is covered with a chemical film (stationary phase). The external surface is coated with polyimide that confers flexibility and robustness to the column.

The stationary phase is characterized by the chemical functions grafted onto the coated silica-based support. The different kinds of columns that are available commercially are too numerous for an exhaustive list. Three types of stationary phases can be distinguished, so the three categories of columns are apolar, low polarity (or semi-polar), and polar. Apolar columns are composed of a chemical film on which methyl groups are fixed to the silica. For the low polarity columns, the polarity is increased by substituting methyl groups with phenyl groups in proportions depending

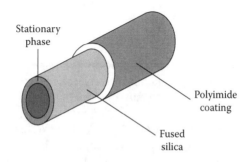

FIGURE 1.7   Capillary column trunk under sagittal sectioning.

on the level of polarity (5 to 30% of phenyl groups in general). The greater the proportion of phenyl groups, the higher the polarity in the capillary column becomes.

The stationary phase of a polar column is composed usually of polar components such as polyethylene glycols or cyano groups. The most used columns are by far the low polarity ones, grafted at 95% with methyl groups and 5% with phenyl groups.

The compounds of a mixture are mainly separated according to their polarities when the stationary phase is polar and according to their volatility when the phase is non-polar. When faced with a mixed mixture, the choice of a low polarity column is usual because the low polarity stationary phases are generally more robust and more thermally stable than their polar counterparts. GC-MS imposes the uses of low bleeding capillary columns (see Section 1.2.2.2).

Unless one uses a post-column made of deactivated silica to limit pressure loss problems, changing a capillary column generally requires putting the spectrometer under atmospheric pressure and therefore stopping the pumping system. Finding a compatible vacuum level for optimal usage of the spectrometer requires waiting for several hours (generally 6 to 12, depending on the device). For this reason, users of GC-MS coupling avoid changing the capillary column unless it is absolutely necessary.

It is possible to place a deactivated silica capillary in the transfer line (Section 1.2.3) and to connect the analytical column to the latter. Obstructing this capillary by jabbing it into a septum when changing the column avoids stopping the pumps of the mass spectrometer. Be aware though, as mentioned previously, all press-fit connections constitute potential sources of leakage.

1.2.2.2   Capillary Column Conditioning

Conditioning a capillary column consists of eliminating the excess of stationary phase introduced in the column during its fabrication. Conditioning is an absolutely essential step in order to obtain efficient chromatographic separations. The stationary phase is relatively fragile; beyond a certain temperature, it deteriorates. Chemical groups detach themselves from the film and are led by the carrier gas (bleeding). These molecules are detected by the detector and produce an increase in background noise in the corresponding chromatogram. The main consequences of bleeding are an increase in the detection limits of the method, wear of the detector, and the pollution of the mass spectrometer source.

Conditioning the column consists in provoking through heating a degree of deliberate wear by eliminating the excess stationary phase. Note that bleeding is a normal phenomenon observed during all chromatographic separations when the column is brought to a high temperature, even after conditioning. The function of conditioning is more to limit than to eradicate the bleeding, whose level must be as low as possible.

The fragility of the column depends on the chemical nature of the stationary phase. The constructor of a column specifies the maximal temperature that can be set for isothermal use and the maximum that can be reached in temperature programming. As an example, most non-polar or low polarity columns have a maximal usage temperature of 320°C in isothermal use and 350°C in temperature programming.

Two conditioning modes are commonly practiced. The first consists of maintaining the column at its maximal temperature for a few hours in isothermal use. The second is subjecting it to several heating cycles: ten cycles from 70 to 350°C, with a temperature gradient of 10°C/min, for example. The second mode is generally more efficient insofar as the intensity of the bleeding depends on the temperature and also on the thermal gradient (speed at which the column is heated). The simplest solution is to apply the instructions given by the manufacturer.

1.2.2.3   Geometrical Parameters of a Capillary Column

Independently of the nature of the stationary phase, the capillary column is characterized by three geometrical parameters: its length (5 to 100 m), its internal diameter (0.1 to 0.5 mm), and the width of its stationary phase (0.1 to 5.0 µm). Each one of these parameters has a determining influence on the quality of the separation.

The analyst has three main objectives: to obtain chromatographic peaks that are as thin as possible (high efficiency), as well separated as possible (high resolution), and minimal analysis time. Resolution is directly linked to efficiency and separation. The efficiency of the column depends on the dispersion of the solute molecules around their retention times: the better the efficiency, the thinner the chromatographic peaks will be. The quality of separation also depends on the relative retention times of the different analytes in the mixture. The resolution increases with the length of the column, at the expense of the analysis time and efficiency.

A 15-m column suffices to separate simple mixtures, but the analysis of complex mixtures requires a column equal to or longer than 30 m. Most applications of GC-MS use columns ranging from 15 to 60 m in length. Increasing the width of the stationary phase multiplies its interactions with the solutes, which improves the chromatographic separation but increases the elution time and, of course, the analysis duration. One generally chooses an important width for the stationary phase for separating very volatile compounds. The increase in internal diameter allows the shortening of the analysis time but leads to a decrease in efficiency.

1.2.2.4   Oven Temperature Programming

The capillary column is placed in an oven because the interactions between compounds and the stationary phase depend on the temperature. Increasing the stationary phase speeds the elution of the compounds and therefore decreases the time of analysis. Most users work with temperature programming. Certain screening methods aim to detect a large number of compounds in a single analysis and frequently use temperature programming ranges from 40 to 350°C.

In practice, when performing an analysis on matrix extracts, one must keep in mind that temperature programming must take into account the volatilities of analytes and also those of the interfering compounds from the matrix that must be eluted to prevent pollution of the capillary column. Since the precise boiling points of interfering compounds are generally unknown, the final programmed temperature of an oven frequently corresponds to the maximal temperature that the column can stand in order to evacuate as many pollutants as possible.

The analysis of very volatile compounds, of certain solvents, for example, can lead to maintaining an oven at a temperature below ambient. In order to do this, a cryogenic liquid, generally carbon dioxide, is sequentially injected into the oven to cool it. The CO_2 is compressed in a cylinder at a pressure around 200 bars. The adiabatic expansion of the gas at its exit from the cylinder allows the cooling. Just like the oven, the injector can be cooled by a cryogenic fluid. Independent of the need to work at a low temperature, cryogenic cooling can be used to cool the oven (and possibly the injector if it is temperature programmable) between two chromatographic runs to diminish the lapse of time between two consecutive injections and increase the number of achievable tests.

In the context of GC-MS coupling, many users prioritize the analysis time at the expense of chromatographic separation because the coelution of analytes is generally not a problem. The reason is that each chromatographic peak is "integrated" on the current of a characteristic ion of the analyte (refer to Chapter 7 covering quantification by GC-MS). It is now very rare to develop GC-MS methods that imply complex temperature programming and are segmented into different thermal gradients. This type of programming that was widespread when conventional detectors were used is obsolete with a mass spectrometer unless a mixture to be separated is very complex. Dosing becomes less precise when one simultaneously quantifies too many co-eluted compounds.

1.2.3 TRANSFER LINE

The capillary column exits the chromatograph oven and enters the mass spectrometer via a transfer line (Figure 1.8) that consists of an intensely heated cylinder (250 to 300°C) for preventing condensation of the eluted molecules between the two devices. In theory, the line transfer temperature, generally fixed in the device's piloting software independently of the programming of the acquisition method, should not be inferior to the final temperature programmed for the oven. However, if the final temperature of the oven exceeds that of the transfer by a few degrees, the difference would not cause any problems since the vacuum present in the mass spectrometer diminishes the boiling points of the analytes when compared to these temperatures under atmospheric pressure. The vacuum therefore contributes to reducing if not stopping the condensation of the analytes.

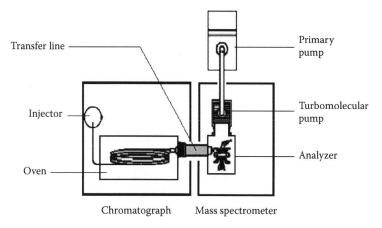

FIGURE 1.8   Representation of a GC-MS device.

The vacuum in the mass spectrometer also influences the retention times of the molecules. The gas is drawn in under the effect of depression (see Section 2.2.3 covering the effects of low pressure).

1.3   FAST CHROMATOGRAPHY

1.3.1   DEFINITIONS

As its name indicates, fast chromatography is a rapid technique utilizing methods with very short analysis times. The point of this technique is to allow high speed to treat many samples in a far shorter time than that required by traditional GC. This is particularly interesting when the GC analysis constitutes a limiting factor for the global analytical chain in terms of duration. It is also very practical for urgent analyses, for example, performing quality control in a factory production chain. The concept of fast GC dates back to the 1960s and interest in this technique grew rapidly in the 1990s when the numbers of samples to be analyzed increased greatly. In a well documented article, E. Matisovà and M. Dömötörovà present a complete description of this technique and its main applications in trace analyzing.[12]

Specialists distinguish three kinds of faster GC techniques: fast, very fast, and ultra fast. Their characteristics are presented in Table 1.1.[13] The ultra fast GC is difficult to perform and imposes many constraints. The main sectors that use fast chromatography are the same ones that use traditional GC, for example for analyzing organic micropollutants and essential oils. As Table 1.1 demonstrates, the time gains can be spectacular.

1.3.2   OPERATIONAL SETTINGS

The seven ways to reduce the length of a chromatographic analysis[14] are:

1. Increasing the flow of carrier gas in the capillary column
2. Increasing the temperature gradient of the oven
3. Reducing the length of the capillary column
4. Reducing the internal diameter of the column
5. Reducing the width of the stationary phase film
6. Using a carrier gas faster than helium: hydrogen
7. Using a detector that works under vacuum

TABLE 1.1
Three Types of Faster GC and Their Main Characteristics

Technique	Analysis Time	Peak Width at Half Height
Fast GC	Minutes	1–3 s
Very fast GC	Seconds	30–200 ms
Ultra fast GC	Less than 1 second	5–30 ms

Several of these possibilities are generally combined to achieve a spectacular decrease in analysis time. In GC-MS coupling, possibility 6 is almost never used. Possibility 7 is systematically applied insofar as the mass spectrometer works under vacuum (refer to Chapter 2). The two first options cannot be used systematically; they depend on the capacity of the chromatograph used.

How can one pass from traditional to fast GC? The question involves more than reducing the length of the capillary column by half in order to halve the elution time. By operating in such a way, one indeed decreases the elution time but also the separation power of the column. This example shows that one must try to adjust different parameters to reduce the analysis time while conserving thin and well-separated chromatographic peaks.

Most applications developed in fast GC use capillary columns of inferior diameter and length than those classically used in GC. To maintain a constant resolution, the length and internal diameter must be reduced in the same proportions. For instance, a method initially developed on a 25-m long and 250-μm diameter column can in fact be replaced by a method using a 10-m long and 100-μm column with no loss in separation power (with an identical stationary phase, of course). Classically, columns used in fast GC present geometrical parameters with the following characteristics: (1) length between 1 and 15 m, (2) internal diameter between 0.10 (mostly) and 0.25 mm, and (3) a stationary phase width ranging from 0.1 to 0.5 μm.

Why don't all laboratories systematically use fast GC because it produces substantial time gains in comparison to traditional GC? Because when reducing the geometrical parameters of the capillary column as suggested above, the capacity of the column is also reduced and this means an increase in the detection threshold of the method. Efficient programs exist to assist the development of a fast GC method from an optimized traditional method.[15,16]

1.3.3 Fast GC Devices

If certain chromatographs are specially designed for fast GC, many late model chromatographs conceived for "traditional" GC also work in fast GC. As seen above, accelerating the analysis can be done by changing the capillary column and also by accelerating the carrier gas flow and/or the temperature gradient of the oven.[17,18] For this, a chromatograph must be equipped with an electronic flow regulator capable of quickly adjusting the column head pressure and bringing it to values higher than 10 bars. The oven must be able to reach very high temperature gradients, and at the same time guarantee a fast cooling process. Flash chromatography therefore uses a system in which the column is placed in a metallic tube heated by electrical resistance at a speed that can reach 1200°C/min.[19,20]

As far as the introduction of the sample is concerned, the classical injection modes (split, splitless, and on-column) are compatible with fast GC provided one adapts the sample quantity to the column's capacity. The main detectors used in gas chromatography (FID, NPD, ECD) are compatible with fast GC. The mass spectrometer is adapted to detection in fast GC and its selectivity is an asset when the analyte retention times and those of the matrix interferents move closer.

However, it is necessary to be vigilant in the choice of the analyzer because not all of them allow high mass spectra acquisition frequencies (refer to Chapter 4 dedicated to analyzers). Indeed, the decrease of the width of the chromatographic peaks (Table 1.1) in relation to conventional GC requires working with mass spectra acquisition frequencies superior to those that are generally used. It is in fact impossible to use GC-MS methods using classical frequency values, corresponding to the recording of 1 to 5 spectra/second when the average width at half-height of the chromatographic peaks is 2 seconds if one wishes to obtain well traced and Gaussian peaks. In this context, the time-of-flight constitutes a quality analyzer for detection in fast GC for it allows recording up to 500 mass spectra/second in scanning mode (Section 4.5 in Chapter 4 discusses time-of-flight analyzers).[21]

1.4   MULTIDIMENSIONAL CHROMATOGRAPHY

1.4.1   COMPREHENSIVE GC (GC×GC)

Comprehensive GC, more frequently known as GC×GC or 2D-GC, is designed for the analysis of particularly complex mixtures that cannot be separated satisfactorily by traditional GC. It utilizes an orthogonal two-column system. In the most frequently used systems, a module placed at the exit of the first column traps the eluted analytes at regular intervals of a few seconds (Figure 1.9).[22] This trapping is ensured by the pulsed dispatch of cryogenic fluid on the capillary column. The fractions thereby obtained are sequentially eluted in the second column following the fast heating of the trap.

The first column is generally a low polar capillary type with classical dimensions, 30 m × 0.25 mm × 0.25 µm, for instance. The second column is inevitably short (1 m, for example) because the elution time of the compounds of a given fraction must not exceed a few seconds since the different trapped fractions follow each other in time intervals of the same order. The second column is generally of higher polarity than the first one.

The chromatograms of GC×GC are radically different from those in traditional GC because the compounds are no longer represented by peaks but by stains in a

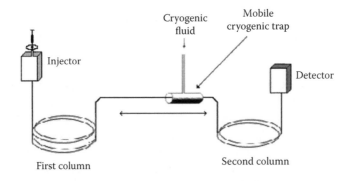

FIGURE 1.9   Classical assembly scheme for comprehensive GC.

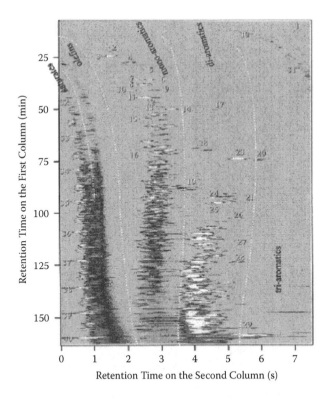

FIGURE 1.10 GC×GC chromatogram of a crude oil distillation fraction. (*Source:* Phillips, J. B. and J. Beens. 1999. *J. Chromatogr. A* 856: 331–347. With permission.)

two-dimensional diagram on which their positions depend on their retention times in each of the two columns. As an example, Figure 1.10 shows a chromatogram obtained by GC×GC after the analysis of a crude oil distillation fraction. The retention times on the first column (0 to 160 min) are on the ordinate axis and the retention times on the second column (0 to 7 sec) are on the abscissa axis.

This technique offers excellent chromatographic resolution and very low detection thresholds.[23] It is generally coupled with mass spectrometry. Spectrometers using time-of-flight analyzers (refer to Chapter 4) are particularly efficient in this context.

1.4.2 Two-Dimensional Chromatography (GC-GC)

Two-dimensional chromatography or GC-GC uses a two-column system as does GC×GC. The two-column configuration is often a source of confusion about the two techniques for non-specialists.

In two-dimensional chromatography, the two columns are long (30 m each, for instance) and situated in different ovens. In the most classical configuration based on the principle known as "heart cutting," the first column is connected to a FID detector, and the second, of different polarity, is connected to a mass spectrometer (rarely to another FID). After injection in the first column, we can visualize

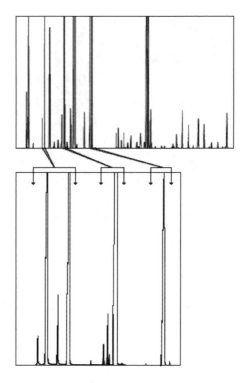

FIGURE 1.11   Principle of heart cutting.

the obtained chromatogram. It is then possible to select a time interval designed to improve the chromatographic separation. During the second injection, the system is programmed for the eluted molecules to be routed via a modulator toward the second column that will allow their separation. Several fractions can be performed during a single GC-GC analysis.

Heart cutting is illustrated in Figure 1.11. The top chromatogram represents the separation on the first column. The bottom one shows three zones corresponding to three fraction dispatches in the second column. The compounds selected are separated correctly in the second chromatogram whereas they were co-eluted in the first one.

Contrary to comprehensive GC, two-dimensional chromatography only leads part of the mixture toward the second column. All the mixture is not treated in two dimensions but the separation of the portion of interest is much better achieved than in GC×GC because it uses a better resolving column.

1.4.3   APPLICATIONS OF MULTIDIMENSIONAL CHROMATOGRAPHY

Multidimensional chromatography techniques (GC×GC and GC-GC) are used when the mixtures to be separated are very complex because of a high number of analytes (hydrocarbons, essential oils, pesticides, etc.) or because of the number of interfering compounds from the matrix susceptible to coelution with the molecules of interest (e.g., environmental analyses, research of metabolites in biological matrices). The main sectors that utilize these methods are petrochemistry,[24] environmental protection,[25] and the perfume and aroma industries.[26,27] In addition to improving resolution greatly, multidimensional chromatography allows in some cases very low detection thresholds that can reach parts per billion (ppb).[28]

For a more exhaustive view on multidimensional chromatography and its usage, one can refer to the synthesis articles by Marriott and Shellie (2002), Dallüge, Beens, and Brinkman (2003), and Mondello et al. (2008).[29–31]

REFERENCES

1. Grob, R. L. 2004. Theory of gas chromatography. In *Modern Practice of Gas Chromatography*, 4th ed., Grob, R. L. and E. F. Barry, Eds. Hoboken, NJ: John Wiley & Sons.
2. Björklund, J., P. Tollbäck, C. Hiärne et al. 2004. Influence of the injection technique and the column system on gas chromatographic determination of polybrominated diphenyl ethers. *J. Chromatogr. A* 1041: 201–220.
3. Cole, R. B. 1997. *Electrospray Ionization Mass Spectrometry: Fundamentals, Instrumentation, and Applications*. New York: Wiley-Interscience.
4. Lesueur, C., P. Knittl, M. Gartner et al. 2008. Analysis of 140 pesticides from conventional farming foodstuff samples after extraction with the modified QuEChERS method. *Food Control.* 19: 906–914.
5. Pang, G. F., Y. M. Liu, C. L. Fan et al. 2006. Simultaneous determination of 405 pesticide residues in grain by accelerated solvent extraction then gas chromatography–mass spectrometry or liquid chromatography–tandem mass spectrometry. *Anal. Bioanal. Chem.* 384: 1366–1408.
6. Chaves, A., D. Shea, and D. Danehower. 2008. Analysis of chlorothalonil and degradation products in soil and water by GC/MS and LC/MS. *Chemosphere* 71: 629–638.
7. Nunez, O., E. Moyano, and M. T. Galceran. 2005. LC-MS/MS anaylis of organic toxics in food. *Trends Anal. Chem.* 24: 683–703.
8. Joice, J. R., T. S. Bal, R. E. Ardrey et al. 1984. The decomposition of benzodiazepines during analysis by capillary gas chromatography/mass spectrometry. *Biomed. Mass Spectrom.* 11: 284–289.
9. Pirnay, S., I. Ricordel, D. Libong et al. 2002. Sensitive method for the detection of 22 benzodiazepines by gas chromatography–ion trap tandem mass spectrometry. *J. Chromatogr. A* 954: 235–245.
10. Grob, R. L., Ed. 1995. *Modern Practice of Gas Chromatography,* 3rd ed. New York: Wiley-Interscience.
11. March, J. and M. B. Smith. 2001. *March's Advanced Organic Chemistry: Reactions, Mechanisms, and Structure*. New York: Wiley-Interscience.
12. Matisovà, E. and M. Dömötörovà. 2003. Fast gas chromatography and its use in trace analysis. *J. Chromatogr. A* 1000: 199–221.
13. Koritar, P. H. G. Janssen, E. Matisovà et al. 2002. Practical fast gas chromatography: methods, instrumentation, and applications. *Trends Anal. Chem.* 21: 558–572.
14. Klee, M. S. and L. M. Blumberg. 2002. Theoretical and practical aspects of fast gas chromatography and method translation. *J. Chromatogr. Sci.* 40: 234–247.
15. Hada, M., M. Takino, T. Yamagami et al. 2000. Trace analysis of pesticide residues in water by high-speed narrow-bore capillary gas chromatography–mass spectrometry programmable temperature vaporizer. *J. Chromatogr. A* 874: 81–90.

16. Sandra, P. and F. David. 2002. High-throughput capillary gas chromatography for the determination of polychlorinated biphenyls and fatty acid methyl esters in food samples. *J. Chromatogr. Sci.* 40: 248–253.

17. Matisovà, E. and M. Dömötörövà. 2003. Fast gas chromatography and its use in trace analysis. *J. Chromatogr. A* 1000: 199–221.

18. Sacks, R., H. Smith, and M. Nowak. 1998. High-speed gas chromatography. *Anal. Chem.* 70: 29–37.

19. MacDonald, S. J., G. B. Jarvis, and D. B. Wheeler. 1998. Ultra fast GC systems using directly resistively heated capillary columns. *Int. Environ. Technol.* 8: 30–32.

20. van Deursen, M. M., J. Beens, H. G. Janssen et al. 1999. Possibilities and limitations of fast temperature programming as a route toward fast GC. *J. High Resolut. Chromatogr.* 22: 509–513.

21. van Deursen, M. M., J. Beens, H. G. Janssen et al. 2000. Evaluation of time-of-flight mass spectrometric detection for fast gas chromatography. *J. Chromatogr. A* 878: 205–213.

22. Phillips, J. B. and J. Beens. 1999. Comprehensive two-dimensional gas chromatography: a hyphenated method with strong coupling between the two dimensions. *J. Chromatogr. A* 856: 331–347.

23. Liu, Z. and J. B. Phillips. 1991. Comprehensive two-dimensional gas chromatography using an on-column thermal modulator interface. *J. Chromatogr. Sci.* 29: 227–231.

24. Schoenmakers, P. J., J. L. M. M. Oomen, J. Blomberg et al. 2000. Comparison of comprehensive two-dimensional gas chromatography and gas chromatography-mass spectrometry for the characterization of complex hydrocarbon mixtures. *J. Chromatogr. A* 892: 229–246.

25. Eljarrat, E. and D. Barceló. 2006. Quantitative analysis of polychlorinated *n*-alkanes in environmental samples. *Trends Anal. Chem.* 25: 421–434.

26. Mondello, L., A. Casilli, P. Q. Tranchida et al. 2005. Comprehensive two-dimensional gas chromatography in combination with rapid scanning quadrupole mass spectrometry in perfume analysis. *J. Chromatogr. A* 1067: 235–243.

27. Dunn, M. S., N. Vulic, R. A. Shellie et al. 2006. Targeted multidimensional gas chromatography for the quantitative analysis of suspected allergens in fragrance products. *J. Chromatogr. A* 1130: 122–129.

28. Ligon, W. V. Jr. and R. J. May. 1984. Target compound analysis by two-dimensional gas chromatography-mass spectrometry. *J. Chromatogr. A* 294: 77–86.

29. Marriott, P. and R. Shellie. 2002. Principles and applications of comprehensive two-dimensional gas chromatography. *Trends Anal. Chem.* 21: 573–583.

30. Dallüge, J., J. Beens, and U. A. Th. Brinkman. 2003. Comprehensive two-dimensional gas chromatography: a powerful and versatile analytical tool. *J. Chromatogr. A* 1000: 69–108.

31. Mondello, L., P. Q. Trancka, P. Dugo et al. 2008. Comprehensive two-dimensional gas chromatography–mass spectrometry: a review. *Mass Spectrom. Rev.* 27: 101–124.

2 Introduction to Mass Spectrometry
General Points

2.1 MASS SPECTROMETER COMPONENTS

Mass spectrometers all contain at least five elements: an analyte introduction system, a source, an analyzer, a detector, and an acquisition and data treatment system.[1–3] Figure 2.1 represents a quadrupole mass spectrometer.

The analyte introduction system is a gas chromatograph. Direct introduction of the sample into the source is rare nowadays because of the difficulty of insertion due to the different pressures between the inside of the source and the exterior. Furthermore, direct introduction is theoretically confined to the analysis of pure compounds that are rare in analytical chemistry. Direct insertion probes are not presented in this book because they do not fall within the scope of GC-MS coupling.

The source is where gaseous ions are produced. The different types of sources used in GC-MS coupling and the various ionization modes and their characteristics are presented in Chapter 3.

The analyzer is the part of the mass spectrometer that allows the separation of ions. The different types of analyzers, their principles of operation, and their main technical features are presented in Chapter 4. The various programmable acquisition modes for analyzers and their respective appeals are detailed in Chapter 5. Chapter 6 is dedicated to a detailed comparison of the performances of quadrupolar analyzers. It should be noted that in certain mass spectrometers like ion traps using internal ionization, the source and the analyzer constitute a single element that plays both roles.

Detectors, as their name indicates, are intended to detect and enumerate ions. There are few kinds of detectors among those used in GC-MS coupling and their mode of operation is simple. For these reasons, there was no purpose in dedicating a chapter to them; they are presented in Section 2.3.

No matter which mass spectrometer is considered, the control system and acquisition and data treatment are handled by personal computers with Windows operating systems. The same computer pilots the chromatograph and the mass spectrometer. Software allowing conversion of formats can be useful for viewing chromatograms and mass spectra acquired on different GC-MS coupling devices.[4]

This chapter covers two subjects that were difficult to position in this book: (1) a vacuum that is essential to the proper functioning of sources, analyzers, and detectors and (2) the operation principles of detectors.

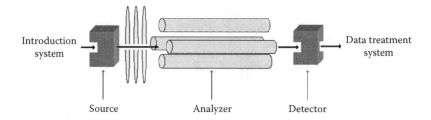

FIGURE 2.1   Quadrupole mass spectrometer.

2.2   VACUUM

2.2.1   WORKING UNDER VACUUM

All mass spectrometers work under vacuum conditions, that is, under a pressure inferior to atmospheric pressure, generally ranging from 10^{-5} to 10^{-7} torr (primary vacuum). A secondary vacuum is characterized by pressures from 10^{-2} to 10^{-3} torr. The quality of the vacuum is a determining factor for all analyses.

The process consists of evacuating the carrier gas arriving in the source of the spectrometer and the residual atmospheric molecules such as nitrogen, oxygen, carbon dioxide, water, and eluted molecules from the chromatograph that have not been ionized and may contribute to pollution of the mass spectrometer. Another goal is to ensure correct functioning of the filament and the electron multiplier of the mass spectrometer that may be weakened by excessive pressure.

In terms of analytical performances, the consequences of an insufficient vacuum are numerous and disastrous. All analyzers use a field (e.g., electromagnetic or magnetic) to separate ions according to their mass-to-charge ratios. The trajectory of the ions in the field must be as precise as possible. If the ions collide with residual molecules, they can react with them, deviate from their trajectory, and be fragmented by the collisions. These phenomena will result in resolution, sensitivity, and spectral reproducibility problems. Furthermore, residual molecules are susceptible to being ionized. The resultant ions will interfere with the characteristic ions of the analytes within the mass spectra.

It is important to underline that a 1-l volume under a 10^{-6} torr pressure still contains about 4.10^{13} molecules—very far from a "total" vacuum. Consequently, it is impossible to avoid all collisions between ions and molecules; at best, one can minimize them.

The mass spectrometist community keeps using different international pressure units, whether they are official or not. In this book, the choice was made to systematically express pressure in torr when it directly concerns mass spectrometry and in bar in all other cases. Table 2.1 shows the conversion factors of different pressure units.[5]

2.2.2   PUMPING SYSTEM

At present, no pump is capable of bringing a chamber to the secondary vacuum required for a spectrometer to function properly. All spectrometers are therefore equipped with at least two pumps: a mechanical pump or a primary pump (Figure 2.2)

TABLE 2.1

Conversion Factors of Pressure Units Used in Mass Spectrometry

	Pascal	bar	kg/cm²	atm
1 Pascal	1	1.10^{-5}	$1.0197.10^{-5}$	$9.8692.10^{-4}$
1 bar	1.10^{5}	1	1.0197	9.8692
1 kg/cm²	$9.8067.10^{4}$	$9.8067.10^{-1}$	1	$9.6784.10^{-1}$
1 atm	$1.0133.10^{5}$	1.0133	1.0333	1
1 torr	$1.3332.10^{2}$	$1.3332.10^{-3}$	$1.3595.10^{-3}$	$1.3158.10^{-3}$
1 mbar	1.10^{2}	1.10^{-3}	$1.0197.10^{-3}$	$9.8692.10^{-4}$
1 mm Hg	$3.386.10^{3}$	$3.386.10^{-2}$	$3.453.10^{-2}$	$3.345.10^{-2}$
1 PSI	$6.8948.10^{3}$	$6.8948.10^{-2}$	$7.0306.10^{-2}$	$6.8046.10^{-2}$

	torr	mbar	1 mm Hg	PSI
1 Pascal	$7.5006.10^{-3}$	1.10^{-2}	$2.953.10^{-4}$	$1.4503.10^{-4}$
1 bar	$7.5006.10^{2}$	1.10^{3}	$2.953.10^{1}$	$1.4503.10^{1}$
1 kg/cm²	$7.3556.10^{2}$	$9.8068.10^{2}$	28.96	14.22
1 atm	760	1013	$2.995.10^{1}$	$1.42247.10^{1}$
1 torr	1	1.3332	$3.937.10^{-2}$	$1.9337.10^{-2}$
1 mbar	$7.5006.10^{-1}$	1	0.02953	$1.4503.10^{-2}$
1 mm Hg	$2.540.10^{1}$	$3.386.10^{1}$	1	$4.910.10^{-1}$
1 PSI	$5.1715.10^{1}$	$6.8947.10^{1}$	2.041	1

FIGURE 2.2 Primary pump.

for ensuring a primary vacuum that is relayed by a turbomolecular pump allowing the system to reach secondary vacuum.

The turbomolecular pump (Figure 2.3) is composed of blades mounted on a rotor capable of several thousands of rotations per minute. The turbomolecular pump does not generally require any maintenance. However, the oil in the primary pump must be regularly de-gassed to eliminate the gas bubbles trapped in the oil that will diminish lubrication. The primary pump oil must also be periodically drained and replaced

FIGURE 2.3 Turbomolecular pump.

every 3 to 12 months according to use. The quality of the oil is a determining point. It must combine good lubrication and low vapor pressure characteristics in order to prevent the oil vapors from being sucked in by the vacuum in the mass spectrometer. It is consequently very important to use the specific primary pump oil designed for this purpose. A primary pump is relatively robust but deficiencies in maintenance can lead to consequences for the secondary pump. Generally, a secondary pump is more fragile and must work harder to supply a satisfactory vacuum if the primary pump is not efficient enough. Dry pumps are recent innovations. They do not use any oil and their maintenance needs are very basic: one must simply clean a filter periodically.

2.2.3 EFFECTS OF LOW PRESSURE

The operation of mass spectrometers under vacuum leads to various consequences. First, vacuum alters the retention times of analytes compared with those obtained under the same chromatographic conditions using a detector of the type FID, NPD, or ECD (refer to Chapter 1 dedicated to gas chromatography). Indeed, these detectors work under atmospheric pressure or slight overpressure. The depression in the mass spectrometer leads to aspiration of the carrier gas and consequently of the analytes that the gas conveys. The aspiration spans several dozens of centimeters from the extremity of the capillary column connected to the mass spectrometer.

The acceleration of elution translates into a decrease of the compound retention times compared to those obtained in FID, for instance. Obviously, the shorter the column is, the more the phenomenon is pronounced. It is not problematic because the elution order of the compounds is conserved. One must nevertheless keep in mind that, when comparing two chromatograms (one in GC, the other in GC-MS), the analyte retention times change even though the chromatographic conditions are identical.

Another consequence of vacuum is of a practical nature. Users of GC-MS generally work with a single capillary column because changing columns generally requires stopping the pumps. Obtaining a secondary vacuum after atmospheric pressurization of the spectrometer generally takes a few hours, an inconvenience that is fairly incompatible with the specifications required by most analysis laboratories.

A deactivated silica capillary connected to the analytical column at the transfer line level can allow a column change without stopping the pumping. There are also valve systems that allow changing the column without pausing the pumping but these systems often cause absorption of the analytes at the level of the valve. One must therefore check, before using this kind of system, that the valve components are passivated to avoid adsorption of the molecules.

2.2.4 DELAY BEFORE DATA ACQUISITION

The acquisition of mass spectra does not generally start at time zero (time of injection) of the chromatographic run. This delay that lasts several minutes, during which the filament and the detector are turned off, corresponds to elution of the injection solvent. One microliter of solvent arriving in the source leads to an increase of pressure that could risk rupture of the filament if it is turned on. This delay also helps preserve the detector from premature damage because the ionization of the solvent would produce millions of ions arriving simultaneously to the detector.

To determine the duration of this delay, it is possible to inject a very small amount of solvent ($\leq$0.1 µL) and lower the current applied to the filament and the voltage applied to the detector to protect these elements. This allows determination of the retention time of the solvent under the chromatographic conditions of the chosen method.

Programming a delay in the acquisition of mass spectra is pertinent only if the injection solvent is the first eluted compound of the mixture to be separated. One must look for another solution to protect the filament and detector when the method is dedicated to the analysis of solvent mixtures (residual solvents, for instance). Solid phase micro extraction (SPME) probably constitutes the best alternative in this case.[6–8]

2.3 DETECTORS

A detector plays a double role: (1) detecting ions proportionally to their numbers and (2) amplifying the corresponding current to a value of 10^{-12} amperes to make it detectable by the system electronics. Indeed, the electronic cards with which the mass spectrometers are equipped require a minimal current of 10^{-9} A to work. Most commercial GC-MS devices are equipped with electron multiplier detectors like those described next. Sections 2.3.3 and 2.3.4 are dedicated to off-axis detection and signal processing.

2.3.1 ELECTRON MULTIPLIERS

The three main kinds of electron multipliers are (1) the discrete dynode electron multipliers, (2) the continuous dynode multipliers (Figure 2.4) that work in a similar way, and (3) microchannel plates that are discussed in Section 2.3.2.[9]

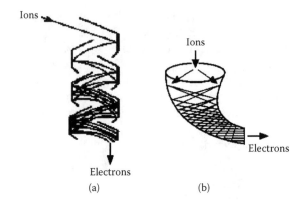

FIGURE 2.4 Electron multipliers with discrete dynodes (a) and continuous dynode (b).

A dynode is a metallic plate covered with an alloy (lead or lead oxide most of the time) that is very rich in electrons. Under the effect of a chock corresponding to the arrival of ions, the alloy emits electrons. In the case of a discrete dynode detector (Figure 2.4a), a difference in potential accelerates these electrons to a second dynode where their arrival provokes the emission of other electrons (more numerous) that are then accelerated to a third dynode and so on. This is referred to as an electron avalanche.[10]

A continuous dynode (or chaneltron) presents in the shape of a curved funnel or a horn of plenty (Figure 2.4b). The interior is covered with an alloy of the same type as the one mentioned above. A difference in potential is applied between the entrance and the exit of the funnel. When an ion hits the internal panel, an emission of electrons is accelerated by the difference in potential and bounces back and forth against the panel due to the shape of the dynode. Each collision detaches new electrons that are also accelerated.[11] A continuous dynode takes up slightly less room than a discrete dynode system.

Discrete dynodes and chaneltrons supply a gain of 10^5; that is, the arrival of an ion is translated by a current emission of about 10^5 electrons. The difference in potential applied (generally ranging from 1000 to 3000 V) must be regularly optimized to make the gain constant. Indeed, the alloy becomes depleted in electrons and the voltage must be increased to compensate for this wear. Gain regulation is automatically done by the mass spectrometer control system; the difference in potential increases over time, following an almost exponential curve (Figure 2.5). The more the dynode surface depletes in electrons, the stronger the acceleration tension must be. The stronger this tension, the more the alloy depletes in electrons.

The electron multiplier is considered the most expensive consumable element of a mass spectrometer. Its life span generally ranges from 10 months for intensive use to 2 to 3 years for a rarely used detector.

2.3.2 Microchannel Plates

The operational principles of microchannel plates (MCPs) are very close to the operational principles of electron multipliers described previously. An MCP (Figure 2.6) is

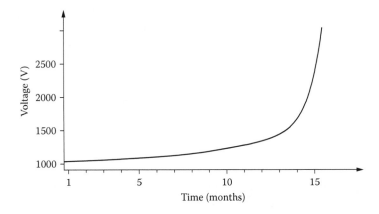

FIGURE 2.5   Voltage applied to electron multiplier as a function of time.

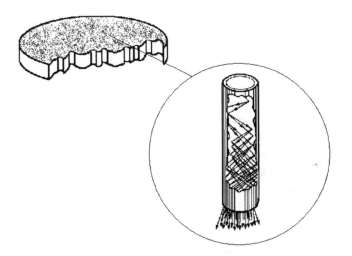

FIGURE 2.6   Microchannel plate and zoom on a microchannel.

composed of thousands of microchannels. Each microchannel plays the role of a small electron multiplier. The inner surface of the channel is covered with a coating rich in electrons. Electrons are emitted when an ion collides with the surface of the plate.

A difference in potential applied between the two faces of the plate accelerates the emitted electrons. The collision with the surfaces leads to the emission of more electrons, allowing amplification of the current. The gain supplied by a microchannel plate is generally of a 10^3 value, which is insufficient to treat electronically the electrical current caused by the impacts of ions. Consequently, the detector for a mass spectrometer generally consists of at least two microchannel plates assembled in series.

The main advantage of this kind of detector is that it presents an ion collection surface superior to those of the detectors presented previously (several dozens of square

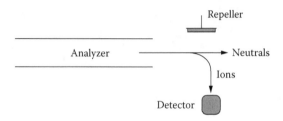

FIGURE 2.7   Principle of off-axis detection of ions.

centimeters compared to approximately 1 cm² for a chaneltron).[12] It is therefore a detector that may be adapted to time-of-flight analyzers within which the ion beam is much larger than in other analyzers (refer to Chapter 4 dedicated to analyzers).

2.3.3   OFF-AXIS DETECTORS

On some mass spectrometers, detection is undergone off axis—the detector is not positioned in the axis on which the ions exit the analyzer. In this case, a repeller diverts the ions at their exit of the analyzer to orient their trajectory toward the detector (Figure 2.7). The repeller is an electrode lead to a positive potential to push back the positive ions or to a negative potential to push back the negative ions.

Positioning the detector off the exit axis of the analyzer prevents neutrals from striking the detector. The arrival of neutrals at the detector increases the background noise in mass spectra and therefore also in the corresponding chromatograms. How can neutrals be emitted from the analyzer when only ions are accelerated from the source by the lenses (refer to Chapter 4)? The presence of neutrals at the exit of the analyzer is attributable to the dissociation of ions referred to as metastable. These ions fragment themselves after leaving the source, either on their trajectory between the source and the analyzer or inside the analyzer. The neutrals resulting from the fragmentation of metastable ions pursue their course in the ions' direction at the time of fragmentation.

The main function of off-axis detection is to reduce as much as possible the background noise in mass spectra. Furthermore, if the potential set on the deflector is done to accelerate the ions before they collide with the detector; it contributes to increasing the gain.

2.3.4   SIGNAL PROCESSING

The arrival of a group of ions at the detector translates as a signal in the shape of a Gauss curve. The Gauss curve is as thin as the arrivals of ions are simultaneous. For a better reading of mass spectra, the signal is integrated so as to be represented in a bar graph form. Each bar corresponds to a mass-to-charge ratio (m/z) and the length of the bar indicates the relative abundance of the ion (Figure 2.8). If the initial signal is too large due to time dispersion at the arrival of the ions of same mass-to-charge ratio (Chapter 4) or saturation of the detector, the integration of the peak generates several bars.

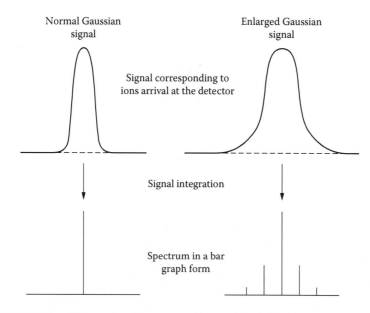

FIGURE 2.8 Signal integration for a group of ions arriving at the detector.

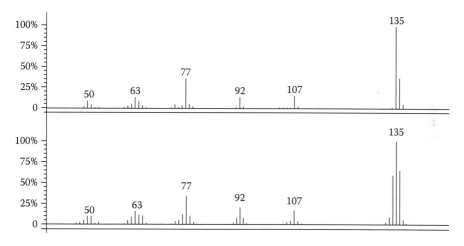

FIGURE 2.9 Mass spectra of cinnamic aldehyde. Top: normal spectrum. Bottom: spectrum recorded with a saturated detector.

An attentive observation of a mass spectrum can therefore reveal problems in the settings, pollution of the spectrometer, or saturation within the detector, as shown in Figure 2.9. The figure compares a mass spectrum of normal aspect (top) with a mass spectrum of the same compound under conditions in which the detector was saturated (bottom) with too many ions arriving simultaneously due to the injection of an overly concentrated solution in analytes. For the reasons stated above, the saturation of the detector translates by producing a spectrum presenting grouped peaks in the geometrical shape of a pine tree.

REFERENCES

1. de Hoffmann, E. and V. Stroobant. 2007. *Mass Spectrometry: Principles and Applications,* 3rd ed. Chichester: John Wiley & Sons.
2. Gross, J. H. 2011. *Mass Spectrometry: A Textbook,* 2nd ed. Berlin: Springer.
3. Watson, J. T. and O. D. Sparkman. 2007. *Introduction to Mass Spectrometry: Instrumentation, Applications, and Strategies for Data Interpretation,* 4th ed. Chichester: John Wiley & Sons.
4. Scientific Instrument Services. 2012. http://www.sisweb.com/software/masstransit.htm
5. OnlineConversion.com. 2010. http://www.onlineconversion.com/pressure.htm
6. Pawliszyn, J. 1997. *Solid Phase Microextraction: Theory and Practice.* New York: Wiley-VCH.
7. Ouyang, G. and J. Pawliszyn. 2006. SPME in environmental analysis, *Anal. Bioanal. Chem.* 386: 1059–1073.
8. Dugay, J., C. Miège, and M. C. Hennion. 1998. Effect of the various parameters governing solid-phase microextraction for the trace determination of pesticides in water. *J. Chromatogr. A* 795: 27–42.
9. Knoll, G. F. 2000. *Radiation Detection and Measurement,* 3rd ed. New York: John Wiley & Sons.
10. Khandpur, R. S. 2006. *Handbook of Analytical Instruments,* 2nd ed. New Delhi: Tata McGraw-Hill.
11. Burroughs, E. G. 1969. Collection efficiency of continuous dynode electron multiple arrays. *Rev. Sci. Instrum.* 40: 35–37.
12. Ladislas-Wiza, J. 1979. Microchannel plate detectors. *Nucl. Instr. Methods* 162: 587–601.

3 Ion Production in GC-MS Sources

3.1 DEFINITIONS AND GENERAL POINTS

The source of a mass spectrometer is where the gaseous ions are produced from the introduced molecules. The nature of the used source depends on the physical state of the substance to be analyzed.[1] One can therefore use an ionization–desorption source when the analyte is a solid substance and an ionization–desolvation source when it is a liquid.[2–5] In coupling with a gas chromatograph in which the eluted compounds arrive into the mass spectrometer in gaseous form, the sources that are used are said to be in electron ionization (EI) or chemical ionization (CI).[6,7] Their use is reserved for the analysis of gaseous or easily volatizable compounds (boiling points not exceeding 400°C). The source is maintained at a high temperature (generally ranging from 100 to 250°C) to avoid the condensation of the analytes. The boiling points and sublimation points of the analytes are, however, much lower under secondary vacuum than under atmospheric pressure.

3.2 ELECTRON IONIZATION

3.2.1 PRINCIPLES

EI consists of submitting molecules to a high energy electron beam. One often speaks of *electron impact* although the term is improper (physicians demonstrated the impossibility of such an impact on M molecules; the energy of the electrons is insufficient for them to penetrate the electronic shells of the molecules). This energy nevertheless allows removal of an electron from the molecule M, leading to the creation of a radical ion $M^{+\cdot}$.

The energy of the incident electrons is such that the $M^{+\cdot}$ ion acquires a great quantity of internal energy (described below) that generally leads to the fragmentation of the ion into smaller daughter ions, also called fragment ions or product ions.

The electronvolt (eV) is an energy unit frequently used by mass spectrometrists. By definition, 1 eV is the kinetic energy of an electron accelerated by 1 V.

The electrons are produced by heating a metallic filament (usually tungsten or rhenium) and accelerated by a voltage of 70 V that provides them a kinetic energy of 70 eV. Disposing of an international standard for the ionization energy allows a comparison of mass spectra recorded on different types of devices. Another benefit is access to databases that classify many thousands of mass spectra, which eliminates the need to interpret the spectra to identify the analytes. The databases are

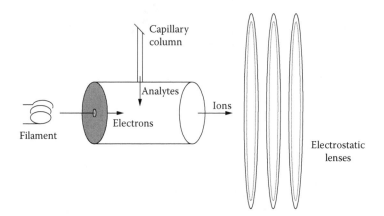

FIGURE 3.1   Representation of a source.

obviously efficient only if they contain the spectra of studied compounds. When the spectra are not included in available databases, the analyst must try to interpret them according to the dissociation pathways presented in Chapter 9. The database details are described in Chapter 8.

Figure 3.1 shows an external source typical of those often used by GC-MS spectrometers. Ions are formed in a metallic cylinder called ion volume whose measurements vary but do not generally exceed a diameter of 1.0 cm for a length of 1.5 cm. This cylinder is open at the extremity that is oriented toward the analyzer; it is pierced with a minimum of two holes, one through which the molecules that exit the capillary column of the chromatograph enter, and the other through which the electrons necessary for EI enter.

An extraction system for the formed ions is located in the prolongation of the source. This extraction system is mostly composed of electrostatic lenses, a hexapole, or a combination of these two systems. The functions of this extraction system are (1) to transfer the ions of the source to the analyzer and (2) to focus the ion beam that is naturally divergent since the entities of same charge mutually repulse each other according to Coulomb's law. The extraction and focalization systems of ions are described in Chapter 4.

In certain mass spectrometers known as internal ionization ion traps the same space within a single device plays in turn the roles of source and analyzer. The principle of electron ionization remains the same but we will see in Chapter 4 dedicated to analyzers, the specificities of internal ionization ion traps compared to systems using an exeternal source.

3.2.2   ION INTERNAL ENERGY

3.2.2.1   Concept

The internal energy of an ion is the energy it acquires following its formation. The energy that must be supplied to detach an electron from the molecule is the ionization potential (IP).[8] It is a thermodynamic parameter intrinsic to a molecule, like its

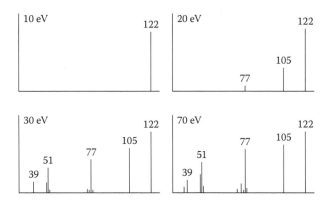

FIGURE 3.2 Mass spectra (simplified) of benzoic acid recorded at various ionization energy values.

boiling point or dipolar moment, for example, that may be determined by experimentation or estimated through molecular orbital calculations. Reference tables are available in the literature.[9]

The aspect of a mass spectrum depends on the internal energy of the formed ions and the internal energy depends on the ionization energy used. As an example, Figure 3.2 represents four mass spectra of benzoic acid obtained with four different values of ionization energy (10, 20, 30, and 70 eV). One can observe that the fragmentation ratio increases logically according to the ionization energy.

The value of 70 eV, which is universally used in electron ionization, is vastly superior to the ionization potentials of organic molecules. The internal energy (E_{int}) corresponds to the difference between the ionization energy that was really supplied to the system and the IP of the molecule. It is an energy that disperses itself essentially under a vibrational form and resembles, to a certain extent, thermal energy. The more important it is, the more the molecular ion is susceptible to fragmentation. A simple thermodynamic diagram of the fragmentation of $M^{+\cdot}$ into a daughter ion $A^{+\cdot}$ and a neutral N is presented in Figure 3.3.

In the figure, TS corresponds to the transition state; it is a specific geometry which the ion $M^{+\cdot}$ must go through to fragment itself. AE corresponds to the activation energy of the fragmentation reaction, that is, the energy that must be supplied

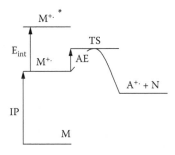

FIGURE 3.3 Thermodynamic diagram of formation and dissociation of $M^{+\cdot}$ ion.

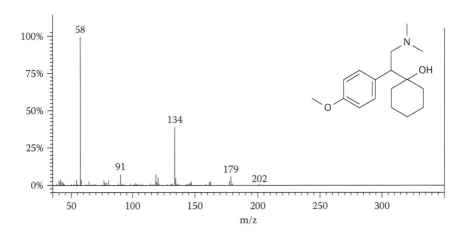

FIGURE 3.4   Electron ionization mass spectrum of venlafaxine (MW = 277).

to the system for it to reach the transition state. An ion can dissociate itself when its internal energy exceeds the activation energy of the corresponding fragmentation. EI is referred to as a "hard" ionization mode because the formed molecular ions have a lot of internal energy. The internal energy leads them to fragment abundantly, so much, in fact, that the molecular ion is frequently absent from the mass spectrum (as in the electron ionization spectrum of venlafaxine in Figure 3.4).

3.2.2.2   Internal Energy Distribution

In electron ionization, internal energy distribution is very important. This means that ions possess very different internal energies among the formed $M^{+\cdot}$ ion population. This phenomenon occurs because (1) all the molecules M do not arrive in the source with the same energy because they clash and collide with the omnipresent helium atoms and residual atmospheric molecules, and (2) all the electrons emitted by the filament do not collide with the molecules with the same kinetic energy (70 eV is the average value). These electrons have different speed characteristics according to the part of the filament that emits them; they are also subject to collisions with helium atoms and H_2O, N_2, and O_2 molecules present in the source.

This important distribution of internal energy explains the complexity of the spectra obtained in electron ionization. The molecular ions $M^{+\cdot}$, when observed in the spectrum, result from species that do not have the sufficient internal energy to fragment themselves. Many fragments otherwise observed result from molecular ions that possess enough internal energy to fragment in different ways.

Fragmentation pathways can be consecutive or competitive. Let us consider the electron ionization spectrum of benzoic acid at 20 eV in Figure 3.2. Three ions are present: the molecular ion $M^{+\cdot}$ at m/z 122 and daughter ions at m/z 105 and m/z 77. Suppose that we are unaware that the compound is benzoic acid. The m/z 77 ion can result from the fragmentation of the m/z 105 ion: the ion at m/z 122 ion fragments into m/z 105, which has enough internal energy to, in turn, fragment into m/z 77. In this case, the fragmentations are consecutive.

The m/z 77 ion could also come from a fragmentation pathway of the m/z 122 ion with no link to that of m/z 105. In this case, the fragmentations m/z 122 → m/z 105 and m/z 122 → m/z 77 are competitive. We will see in Chapter 5 how tandem mass spectrometry establishes *transitions,* i.e., the relationships between the ions of a mass spectrum. The knowledge of such transitions is very useful for mass spectrum interpretation.

3.2.3   IONIZATION YIELD

The ionization yield I of a molecule M can be defined as follows:

$$I = aPVi$$

V is the volume of the source (it is constant), P is the partial pressure in analyte M within the source (P depends on the quantity of analyte introduced), a is a value that depends on the physico-chemical properties of M (IP in particular), and i is the ionization current (current of electrons produced by the heated filament). The ionization yield is variable from one molecule to another. It is considered not to exceed 1/1000 in the best of cases, which means that one forms, at best, 1 ion for 1000 introduced molecules!

Forming so few ions is not really a problem in terms of detection limits because the detectors in mass spectrometers (Chapter 2) are very sensitive. The problem concerns what will become of the non-ionized species. If one neglects gravity, there are essentially two forces present in the source of the spectrometer: (1) the force exerted by the electric field that extracts formed ions and (2) the suction force exerted by the pumping system. The first force exerts itself only on the ions and the second concerns only neutral species in a first approximation, because the force exerted on the ions by pumping is negligible compared to that exerted by the electric field.

If the extraction of ions is quasi-instantaneous since it is undergone in approximately 10^{-6} seconds, the extraction of neutrals is much slower. Therefore, the non-ionized molecules diffuse in the source and will be adsorbed (insofar as they have low volatility) on the panels of the source and on the electrostatic lenses of the extraction system. These neutrals contribute vastly to the pollution of the system and make it compulsory to periodically dismantle the source to cleanse it. When possible, one should be careful not to inject more analytes than necessary as to avoid polluting the source prematurely.

3.2.4   LIMITATIONS OF EI

The universality of EI (all organic molecules are ionizable by the process) justifies the success of this technique, which is by far the most used in GC-MS coupling. It nevertheless happens that this technique is unsuitable for certain analyses. In structural elucidation, the fact that the molecular ion is not always present poses a crucial problem. Indeed, the interpretation of a mass spectrum implies establishing the observed fragmentations from a molecular ion. Confronted with the electron ionization spectrum of an unknown product, it is difficult if not impossible to start the interpretation since one cannot know whether the highest m/z ratio present in the spectrum corresponds to the molecular ion or not. Chemical ionization (see next section) is very valuable in

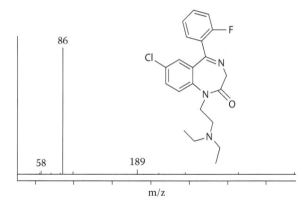

FIGURE 3.5   Electron ionization mass spectrum and chemical formula of flurazepam (MW = 387).

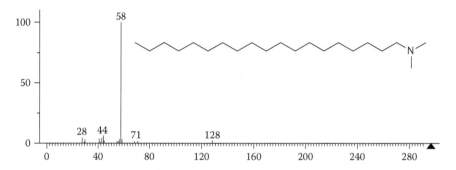

FIGURE 3.6   Electron ionization mass spectrum and chemical formula of N,N-dimethyl-octadecanamine (MW = 297).

this case because it systematically supplies molecular (or pseudomolecular) ions and therefore allows accessing the molecular weight of the analyte.

The electron ionization mass spectrum of flurazepam represented in Figure 3.5 illustrates perfectly the limits of this ionization technique. We can indeed see that the spectrum is dominated by an ion at m/z 86 and that there are no ions of significant abundance at high m/z ratios whereas the molecular weight of the analyte is 387. The m/z 86 ion is not particularly characteristic of flurazepam and it will therefore be impossible to have a specific method of quantification for this compound in EI.

In the same way, the EI mass spectrum of N,N-dimethyloctadecanamine (MW = 297) represented in Figure 3.6 shows the absolute absence of molecular ions. The spectrum is largely dominated by an ion at m/z 58, which is not specific to the molecule.

3.3   CHEMICAL IONIZATION

Chemical ionization (CI) is vastly employed in complement to electron ionization, either for the reasons evoked above (accessing the molecular weight of an analyte), because CI is more selective than EI (all eluted molecules are not ionized, which can constitute an advantage with particularly complex samples), or because CI supplies methods whose detection thresholds are lower than equivalent methods in EI in certain cases.

Compared to EI, CI is a "soft" ionization method. The MH$^+$ ions obtained in CI have much less energy than the M$^+$· ions obtained in EI and fragment themselves much less. CI spectra almost always indicate the molecular weight of the analyte, whereas EI spectra supply more structural information but no certain data on molecular weights. Figure 3.7 compares the EI and CI mass spectra of venlafaxine. The CI mass spectrum shows a majority of MH$^+$ ion; the molecular ion does not appear in the spectrum recorded in EI.

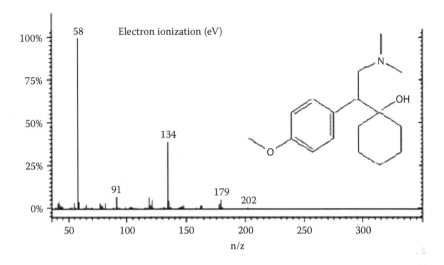

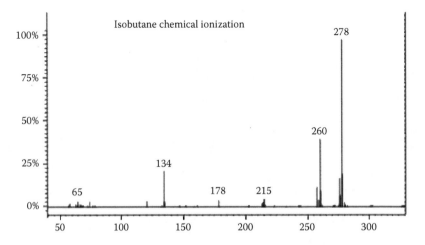

FIGURE 3.7   Mass spectra of venlafaxine (MW = 277) in electron ionization and in isobutane positive chemical ionization.

3.3.1 Positive Chemical Ionization

3.3.1.1 Principles

In CI, the analyte is not directly ionized by the electrons emitted from the filament. A reagent gas (R) is introduced into the source where it undergoes electron ionization (Reaction (3.1)) under a pressure that makes the $R^{+\cdot}$ ions instantaneously react with the non-ionized R molecules to form RH^+ ions with an even number of electrons (Reaction (3.2)). These RH^+ ions react with the analytes M by transferring a proton to them to form MH^+ ions (Reaction (3.3)).

$$R + \bar{e}\,(70\ eV) \rightarrow R^{+\cdot} + 2\bar{e} \tag{3.1}$$

$$R^{+\cdot} + R \rightarrow RH^+ + [R - H]\cdot \tag{3.2}$$

$$RH^+ + M \rightarrow R + MH^+ \tag{3.3}$$

We will see later that other reactions can be in competition with Reactions (3.2) and (3.3). Reaction (3.3) is possible only if the proton affinity of the analyte (affinity of the molecule for the proton) is higher than that of the reagent gas. By definition, the proton affinity of a molecule M is the variation of enthalpy associated with the reaction $MH^+ \rightarrow M + H^+$ in the gas phase.[10] Table 3.1 taken from Harrison's book supplies proton affinity values for families of common organic compounds.[11]

Three reagent gases are mainly used in positive chemical ionization: methane, isobutane, and ammonia. Their proton affinities (in kilojoules per mol) are: methane,

TABLE 3.1
Proton Affinities of Large Families of Organic Compounds

	Proton Affinity (kJ/mol)		Proton Affinity (kJ/mol)
Alcohols and ethers	~711–853	**Amines**	~853–983
Methanol	757	Methylamine	896
n-Propanol	798	Diethylamine	945
Diethylether	838	Triethylamine	972
Aldehydes and ketones	~711–837	**Thiols, sulfites, and nitriles**	~711–879
Formaldehyde	718	Dimethylsulfide	839
Butyraldehyde	806	Methylnitrile	787
Acetone	823	Ethanethiol	858
Acids and esters	~711–837	**Anilines and pyridines**	~870–962
Formic acid	707	m-Chloroaniline	867
Acetic acid	796	n-Methylaniline	912
Ethyl acetate	840	3.Methylpyridine	937

Source: Harrison, A. G. 1992. *Chemical Ionization Mass Spectrometry*. Boca Raton, FL: CRC Press. With permission.

550.1; isobutane, 818.9; and ammonia, 852.7. The choice of the reagent gas depends on precise chemical criteria. Higher reagent proton affinity makes the transfer of the proton to the analyte difficult. The more difficult the proton transfer, the lower the internal energy of the resulting MH^+ ion and the less it fragments. One will therefore choose a reagent gas with a high proton affinity (NH_3) if the aim is not to fragment the MH^+ ion and use a reagent with a very low proton affinity (CH_4) to dissociate this ion to obtain a richer mass spectrum in terms of structural information.

The formation of MH^+ ions is often in competition with that of adduct ions. These adducts are abundant as the analyte is polar and basic. Adducts such as $[M+CH_3]^+$ and $[M+C_2H_5]^+$ with methane or $[M+C_4H_9]^+$ and $[M+C_3H_3]^+$ with isobutane are frequently observed; they are generally less abundant that MH^+ ions. Ammonia supplies an adduct $[M+NH_4]^+$, which is often very abundant and sometimes even more abundant than the MH^+ ion.

Some reactions can enter in competition with Reactions (3.1) to (3.3) described previously. These reactions are charge transfer (Reaction (3.4)) and the abstraction of a hydride or alkyl carbanion (Reactions (3.5) and (3.6)), the latter mainly observed with methane as the reagent.

$$R^+ \cdot + M \rightarrow R + M^+ \cdot \tag{3.4}$$

$$CH_5^+ + M \rightarrow CH_4 + H_2 + [M - H]^+ \tag{3.5}$$

$$CH_5^+ + M \rightarrow CH_4 + C_nH_{2n+2} + [M - C_nH_{n+1}]^+ \tag{3.6}$$

Reaction (3.4) is generally a minor one; it leads to the observation in the spectrum of an ion of m/z ratio equal to the molecular weight of the analyte. Reactions (3.5) and (3.6) can sometimes be predominant. Figure 3.8 illustrates the fragmentations observed in CI with methane, implying mechanisms of the types (3.5) and (3.6) in the case of n-methylbutanamine.

Figure 3.9 compares four mass spectra of the molecule of allethrin: the spectrum recorded in electron ionization and those recorded in positive chemical ionization with methane, isobutane, and ammonia. The molecular weight of allethrin is 302. The molecular ion is absent in the EI mass spectrum. In the spectrum recorded in methane CI, the pseudo-molecular ion MH^+ at m/z 303 constitutes the base peak. It is accompanied by a m/z 301 ion (m/z = M − 1), which is not very abundant. It results from a hydride abstraction reaction and a m/z 331 ion (m/z = M + 29), which corresponds to an adduct between the molecule of allethrin and $C_2H_5^+$.

Daughter ions are observable in the spectrum. They are abundant in comparison to other mass spectra recorded in chemical ionization but much less numerous than in the electron ionization spectrum. This indicates that CI is a "soft" ionization method compared to EI, even if the reagent has low proton affinity. Note that the fragment ions of the CI spectrum have the same m/z ratios as those recorded in EI; this is frequent but absolutely not systematic (see Chapter 9). The spectrum recorded in isobutane CI displays almost exclusively the MH^+ ion. One can furthermore observe a low abundance ion at m/z 302 ($M^+ \cdot$) resulting from a charge exchange reaction. There are almost no fragment ions. The ammonia chemical ionization spectrum shows no

NH2
CH4, H2
+NH2
CH4---H+
H
H3C
H3C
NH2
2CH4
+NH2
CH4---H+
H
H3C
H
NH2
CH4, C3H8
+NH2
CH4---H+
H
H3C
H
CH3

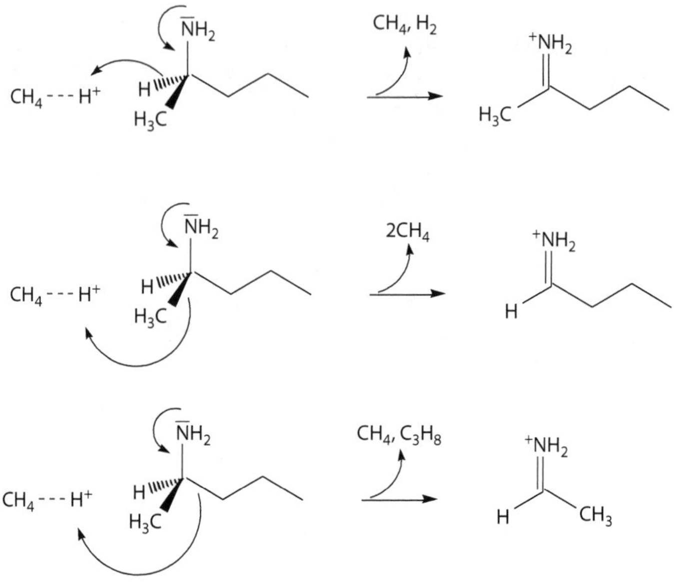

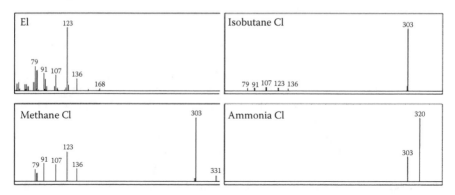

FIGURE 3.8 Mechanisms of hydride and carbanion abstraction from N-methylbutanamine in methane-positive chemical ionization.

EI
123
79
91 107
136
168
Isobutane CI
303
79 91 107 123 136
Methane CI
303
123
79 91 107
136
331
Ammonia CI
320
303

FIGURE 3.9 Mass spectra (simplified) of allethrin (MW = 302) recorded in electron ionization and in positive chemical ionization with methane, isobutene, and ammonia.

daughter ions and the MH^+ ion at m/z 303. It is dominated by a m/z 320 ion (M + 18) resulting from the complexation of allethrin with the ammonium ion NH_4^+.

3.3.1.2 Procedure

For CI to be efficient, the reagent partial pressure in the source must be between 1 and 5 torr. With classic sources described earlier, it is necessary to replace the "open" ion volume used in EI by a "closed" ion volume (see Figure 3.10). Indeed, an open ion volume does not allow attainment of the necessary partial pressure in the reagent because the gas is sucked in by the pumping system at the level of the source. The closed ion volume is not completely closed; it is pierced with a hole to allow the

FIGURE 3.10   Open ion volume (left) for EI and closed ion volume (right) for CI.

exit of the ions. It nevertheless allows the system to reach the pressure necessary for the completion of CI.

The change of ion volume is problematic. Most of the time, it is necessary to "break" the vacuum at the level of the source or even throughout the total mass spectrometer device with all the consequences that this implies (see Chapter 2). Quadrupole users generally make little use of CI for practical reasons. A few recent devices allow the use of semi-open ion volumes, a compromise between open and closed ion volumes. The main advantage is to avoid intervention at the source level when one wants to pass from one ionization mode to another. This type of ion volume is useful for working mainly on structural analyses but does not supply the best results for quantification. We will see below that the ion trap with internal ionization is the most practical device (and often the most efficient) when switching between ionization modes is required.

3.3.1.3   Performing Positive CI Using an Ion Trap with Internal Ionization

In ion trap mass spectrometers using internal ionization (see Chapter 4), the ions are trapped over periods approximately 100.000 times higher than the residency time of the ions in a quadrupole source. With an approximately equal probability of a meeting between ions and neutrals, CI may be performed with a partial pressure in a reagent gas around 100.000 times inferior to that necessary for CI in an external source.[12] This has several beneficial consequences that are detailed below.

No physical intervention is necessary to change the ionization mode. Two clicks of a computer mouse allow switching between ionization modes.

It is possible to change ionization modes during a chromatographic separation. CI is carried out with a partial pressure of reagent so low that it is reached in a few seconds after the opening of the valve that introduces the reagent. In the same manner, returning to EI conditions requires only a few seconds following the closure of this valve. Two compounds can therefore be ionized with different modes within the same chromatogram when their chromatographic peaks are separated by a few seconds.

An application of this property is presented next. An analysis of the 22 main benzodiazepines available in the French market in 1999 raises a problem. Most of the benzodiazepines present lower detection thresholds in EI than in CI but four of them can be analyzed only in CI because they fragment too much in EI. For instance, flurazepam, whose EI spectrum is presented in Figure 3.5, includes a tertiary amino

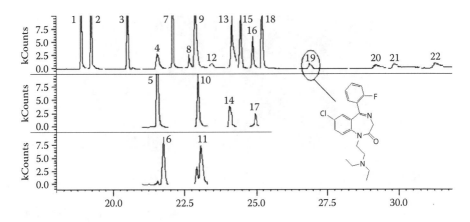

FIGURE 3.11 Chromatogram (recorded in three channels) of a mixture of 22 benzodiaz-epines. Ionization is carried out in CI for compounds 16, 17, 18, and 19, and in EI for all the other compounds. (*Source:* Pirnay, S., Y. Ricordel, D. Libong et al. 2002. *J. Chromatogr. A* 954: 235–245. With permission.)

group that leads it to fragment in EI and provide only the m/z 86 ion ($C_5H_{12}N^+$) in its mass spectrum. One ion is insufficient to characterize an analyte, especially an ion as poorly characteristic as m/z 86 (for a molecular weight of 387.88).

Flurazepam must be analyzed in CI because the internal energy of the pseudo-molecular ions is very low. The ion trap with internal ionization allows analysis of the 22 benzodiazepines in a single injection. Eighteen are ionized in EI and four in CI during the same chromatographic run.[13] Figure 3.11 shows the resulting chromato-gram of this analysis. The chromatogram was recorded in three channels to integrate precisely the peaks of the co-eluted compounds (refer to Chapter 7 dedicated to GC-MS quantification).

Ion traps with internal ionization allow use of a liquid reagent placed in a small flask next to the mass spectrometer. Under the effect of the depression in the spec-trometer, the molecules constituting the vapour pressure at the surface of the liquid are sucked into the ion trap via a small tube. The pressure is easily adjusted by a needle valve. This procedure is both practical and economic. It presents no need for a cylinder of gas under pressure, a two-stage regulator, or a manometer. A few milli-liters of liquid are sufficient to carry out CI for several days to weeks.[14] Methanol and acetonitrile are the two most common liquid reagents; they provide excellent results. Their proton affinities are, respectively, 760.3 and 787.5 kJ/mol.[15]

3.3.2 NEGATIVE CI: ELECTRON ATTACHMENT

In GC-MS, *negative chemical ionization* is an often used but improper term that refers to electron attachment. This ionization mode is generally used for the analy-sis of particularly electrophilic compounds such as certain halides and aromatics—compounds that in "classical" gas chromatography are often detected with a device known as an electron capture detector (ECD).

Electron attachment is described by Reaction (3.7). Although the energy of ionizing electrons is 70 eV in electron ionization, it reaches only 1 to a few electronvolts in electron attachment ("slow" electrons). Consequently, an electron will fix itself to a molecule instead of removing another electron; this leads to the formation of a negative molecular anion with an odd number of electrons.

$$M + \bar{e} \ (slow) \rightarrow M^{\cdot -} \tag{3.7}$$

In practice, it is not possible to accelerate the electrons produced by heating the filament at a few volts. If the acceleration of the electrons is too weak, the filament overheats to breaking. That is why we must accelerate the electrons with a difference of potential up to several dozens of volts (usually 70 V) just as in electron ionization. We first introduce a cooling gas (usually methane or ammonia) in order for the collisions that the electrons will endure with the gas molecules to thermalize the electrons and reduce their kinetic energy considerably. By analogy with positive chemical ionization, we speak of *negative chemical ionization* due to the use of a gas but the reaction does not really play a chemical role in the source. The cooling gas pressure is generally a few torr.

For the reasons specified in the chapter dedicated to analyzers (Section 4.4), it is impossible to proceed to analyses in negative chemical ionization with an ion trap using internal ionization because the thermalized electrons are instantly ejected from the trap.

Figure 3.12 compares two mass spectra of the alprazolam molecule, one recorded in EI and the other in electron attachment. The EI mass spectrum displays many daughter ions while the spectrum recorded under electron attachment exclusively shows the isotopic distribution (see Chapter 9) corresponding to the $M^{\cdot -}$ ions: four

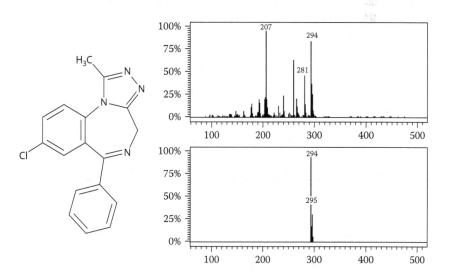

FIGURE 3.12 Mass spectra of alprazolam recorded in EI (top) and in electron attachment (bottom).

peaks corresponding to the isotopomers of chlorine atoms (^{35}Cl and ^{37}Cl) and carbon atoms (^{12}C and ^{13}C). The electron attachment spectrum displays no fragment ions.

3.3.3   DETECTION THRESHOLD AND CI

As seen previously, chemical ionization is frequently used in structural analysis to access the molecular weight of the analyte when electron ionization fragments the molecular ions to the point that they are absent from the EI mass spectrum. One of the less known functions of chemical ionization, whether it is positive or negative, is that it sometimes reduces the detection threshold of an analyte compared to that reached in EI.

Chemical ionization generally supplies fewer ions than electron ionization, in terms of genre and quantity because the internal energy distribution is weaker than in EI. Thus, the recorded ionic current in CI is generally inferior to that recorded in EI chromatograms. Nevertheless, CI can sometimes supply a detection threshold inferior to that of the corresponding EI method. This depends on two factors.

The first factor is that, in terms of detection threshold, one must consider the signal-to-noise ratio and not only the signal associated with the chromatographic peak of the analyte (refer to Chapter 5 dedicated to acquisition modes). Therefore, if the signal recorded in CI is often inferior to that recorded in EI, the noise is almost always lower than in EI, resulting in a better signal-to-noise ratio in this ionization mode.

The second factor is based on the fact that the ionic current is less diluted in chemical ionization.[16] When one develops a quantification method, the corresponding chromatographic peak is not integrated on the total ionic current but on the current corresponding to one or three ions maximum (refer to Chapter 7 dedicated to quantification). In these conditions, the numerous ions present in EI become useless when the analyte is characterized. A large part of the signal is lost when these ions are no longer detected. Conversely, CI supplies very few ions of different m/z ratios in the source (1 to 10 mostly). When one goes from a scanning acquisition mode to a selective detection protocol, the detected ion current stays almost as intense in CI.[17,18]

REFERENCES

1. Gross, M. L. and R. Caprioli. 2006. *The Encyclopedia of Mass Spectrometry*, Vol. 6. Amsterdam: Elsevier.
2. Hoppilliard, Y., Y. Le Beyec, and S. Della-Negra. 1993. Particle-induced desorption-ionization processes for organic and bioorganic molecules: ion formation mechanisms. *J. Chem. Phys.* 90: 1367–1398.
3. Cotter, R. J. 1992. Time-of-flight mass spectrometry for the structural analysis of biological molecules. *Anal. Chem.* 64: 1027–1039.
4. Cole, R. B. 1997. *Electrospray Ionization Mass Spectrometry: Fundamentals, Instrumentation, and Applications*. New York: Wiley-Interscience.
5. Kebarle, P. and U. H. Verkerk. 2009. Electrospray: from ions in solution to ions in the gas phase: what we know now. *Mass Spectrom. Rev.* 28: 898–917.
6. Watson, J. T. and O. D. Sparkman. 2007. *Introduction to Mass Spectrometry: Instrumentation, Applications, and Strategies for Data Interpretation*, 4th ed. Chichester: John Wiley & Sons.

7. Munson, M. S. B. and F. H. Field. 1966. Chemical ionization mass spectrometry. I: General introduction. *J. Am. Chem. Soc.* 88: 2621–2630.
8. Lide, D. R. 2009. *CRC Handbook of Chemistry and Physics,* 90th ed. Boca Raton, FL: CRC Press.
9. Lias, S. G., J. E. Bartmess, J. F. Liebman et al. 1988. Gas-phase ion and neutral thermochemistry, *J. Phys. Chem. Ref. Data* 17.
10. McNaught, A. D. and A. Wilkinson. 1997. *Compendium of Chemical Terminology,* 2nd ed. Oxford: Blackwell Scientific.
11. Harrison, A. G. 1992. *Chemical Ionization Mass Spectrometry.* Boca Raton, FL: CRC Press.
12. Bonner, R. F., G. Lawson, and J. F. J. Todd. 1972. A low-pressure chemical ionisation source: an application of a novel type of ion storage mass spectrometer. *J. Chem. Soc. Chem. Commun.* 11: 1179–1180.
13. Pirnay, S., Y. Ricordel, D. Libong et al. 2002. Sensitive method for the detection of 22 benzodiazepines by gas chromatography–ion trap tandem mass spectrometry. *J. Chromatogr. A* 954: 235–245.
14. Bouchonnet, S., D. Libong, and M. Sablier. 2004. Low pressure chemical ionization in ion trap mass spectrometry. *Eur. J. Mass Spectrom.* 10: 509–521.
15. Bouchonnet, S., S. Kinani, M. Sablier et al. 2007. In situ chemical ionization in ion trap mass spectrometry: the beneficial influence of isobutane as a reagent gas. *Eur. J. Mass Spectrom.* 13: 227–232.
16. Eckenrode, B. A., S. A. McLuckey, and G. L. Glish. 1991. Comparison of electron ionization and chemical ionization sensitivities in an ion trap mass spectrometer. *Int. J. Mass Spectrom. Ion Proc.* 106: 137–157.
17. Ding, W H., J. H. Lo, and S. H. Tzing. 1998. Determination of linear alkylbenzene sulfonates and their degradation products in water samples by gas chromatography with ion-trap mass spectrometry. *J. Chromatogr. A* 818: 270–279.
18. Buchanan M. V., R. L. Hettich, J. H. Xu et al. 1995. Low level detection of chemical agent simulants in meat and milk by ion trap mass spectrometry. *J. Hazard. Mater.* 45: 49–59.

4 Ion Separation
Analyzers

4.1 GENERAL POINTS

4.1.1 CHARGE STATE OF IONS

The ions produced in the source are separated in the analyzer according to their mass-to-charge (m/z) ratios. The Thomson unit (Th, named for a famous mass spectrometrist) corresponds to an m/z ratio of 1; it is widely used in mass spectrometry. In a GC-MS context, we often speak too generally about mass measurement because $z = 1$ in electron ionization and positive chemical ionization and $z = -1$ in negative chemical ionization.

Contrary to what is frequently observed with the electrospray sources used in LC-MS coupling when analyzing large molecules that supply multicharged ions, the ions generated in the sources used in GC-MS coupling are too small to carry several charges. Removing or adding two electrons and adding two protons to a small- or medium-size molecule would require energy that is not available in an EI or CI source. Indeed, in the gas phase, an ion can carry one charge for approximately 1000 mass units. Mass spectra of peptides and proteins recorded in LC-MS display ions illustrating this approximation very well.[1] The GC-MS analysis of compounds with molecular weights greater than 1000 remains an absolute exception; such analysis of compounds with molecular weights reaching 2000 mass units is unthinkable.

Small positively or negatively multicharged ions exist in solution, as in the case of an amino acid in water at an acidic or alkaline pH, for instance. The stability of such a system is ensured by the solvent molecules that surround the ions, especially those of the first shell of solvation that orient themselves according to charge to reduce the global energy of the system. In the gas phase, therefore, in the absence of solvation, the charges are not stabilized. The Coulomb repulsion between two charges of the same kind suffices to dissociate the system.

4.1.2 TYPES OF ANALYZERS

Most of this chapter is devoted to quadrupolar analyzers because they are by far the most common; they constitute more than 90% of the GC-MS devices sold worldwide. Quadrupole analyzers are not the most efficient in all domains because they are low resolution devices. Their commercial success can be explained by their reasonable cost, robustness, versatility, and simplicity of use. The two main categories of quadrupolar analyzers are the quadrupoles (or quadrupolar filters) and the ion traps. As their name implies, the triple quadrupoles consist of three quadrupoles. They are generally used for tandem mass spectrometry and their functioning principles are presented in Chapter 5 dedicated to acquisition modes.

Magnetic sector and time-of-flight (TOF) analyzers are used less frequently in GC-MS coupling; their functioning principles and advantages are described briefly at the end of this chapter.

Ion cyclotron resonance (ICR) analyzers provide very efficient resolution (see below) but are not described in this book because they are very expensive and therefore scarcely used. Moreover, they are almost never coupled to gas chromatography. For information on ICR mass spectrometry, refer to one of the synthesis articles by A. G. Marshall, the "father" of this technique.[2,3]

4.2 TECHNICAL FEATURES

Analyzers are characterized generally by four main technical criteria: mass-to-charge scanning range, transmission, scanning speed, and resolution.

4.2.1 Mass-to-Charge (m/z) Scanning Range

While the range of m/z ratios scanned by most analyzers can reach 10.000 Th or more (the range is theoretically unlimited for a TOF analyzer), it is generally limited to 1000 m/z in GC-MS coupling used to analyze volatile compounds with low to average molecular weights. The m/z range to be scanned is fixed by the user in the acquisition method required by the analytical context. It is generally better not to scan the m/z ratios over a larger range than necessary in the hope of gaining sensitivity as explained in Chapter 5 dedicated to acquisition modes.

4.2.2 Transmission

Transmission is defined as the ratio of the number of detected ions and the number of produced ions; it is reflected by the capacity of the analyzer to lose a minimum of ions between their creation and detection. Transmission is difficult to evaluate because of the inability to know the exact quantity of ions formed in the source. Transmission is never 100% because a significant number of ions is always lost either at analyzer or in the ion transfer zone between the analyzer and the source.

Transmission depends strictly on the operational mode of the analyzer. For instance, the transmission of a quadrupole operated in the selected ion monitoring mode (SIM; see Chapter 5) is much higher than that of the same analyzer in scanning. Transmission is also tightly linked to the detection threshold: the higher the ion transmission, the lower the analyte detection thresholds are.

4.2.3 Scanning Speed

The scanning speed corresponds to the speed at which the analyzer is capable of covering a given range of m/z ratios. For instance, a modern quadrupole has a scanning speed of around 10.000 Thomson/second (Th/s). If one programs an acquisition mode to record the mass spectra between 50 and 549 Th (a range of 500 Th), the mass spectrometer will record 20 spectra per second (10.000:500). The interval between two consecutive points of the chromatographic trace therefore corresponds to 1/20 of a second—a value that allows a very precise chromatographic trace. If the elution of a compound lasts 12 seconds, the corresponding chromatographic peak will be described as $20 \times 12 = 240$ points.

A high scanning speed allows work at a high spectra recording frequency. This parameter is absolutely crucial for quantifying analytes in GC-MS. The precision of the dosage correlates with that of the chromatographic peak of the molecule to be dosed.[4]

4.2.4 Resolution

Resolution traduces the ability of an analyzer to separate ions; it correlates with the precision of the measure of m/z ratios. The analyzers are generally separated into two categories: low and high resolution. The first category groups quadrupoles and ion traps, the second includes ICR, TOF analyzers equipped with electrostatic reflectors, and devices that combine a magnetic sector with an electric sector.

4.2.4.1 Low Resolution

The resolution supplied by a quadrupole analyzer is related to its scanning speed in m/z: the lower the scanning speed, the better the resolution. With an analyzer of this kind, one mostly works at a very high scanning speed and unitary resolution. This means that the m/z ratios of ions are given with no decimal figures. Scanning at a lower speed would generate m/z ratios of only two digits. This is not a big advantage with GC-MS because m/z ratios with two digits do not allow determinations of molecular formulas.

4.2.4.2 High Resolution

One generally speaks of high resolution when an analyzer measures the m/z ratios of the ions with a precision of at least four decimals. High resolution allows distinction of ions known as isobars; that is, they possess the same m/z ratios even though their chemical formulas are different. For instance, the $C_5H_{14}ON^{+\cdot}$ and $C_5H_{16}N_2^{+\cdot}$ ions are isobars with m/z ratios of 104. High resolution allows their separation because the first's m/z ratio is 104.10699 and the second's is 104.13080.

Table 4.1 shows the exact masses of the main isotopes of the chemical elements most frequently encountered in molecules that can be analyzed by GC-MS. To determine the exact m/z ratio of a radical ion $M^{+\cdot}$ to at least four decimals, it is necessary to subtract the mass of the electron detached from the molecule [m(electron) = 0.0005486].

4.2.4.3 High Resolution Applications

High resolution has three main applications. First, it allows determination of the molecular formula of a compound. From a given m/z ratio (to several decimal places), computer software suggests various possible molecular formulas.[5] The better the resolution, the more decimals in the result. This in effect reduces the number

TABLE 4.1

Isotopic Abundances and Exact Masses of Main Isotopes of Elements Most Frequently Encountered in Molecules Analyzed by GC-MS

Chemical Element	Main Isotopes	Isotopic Abundances (%)	Exact Mass
Carbon	^{12}C	98.90	12.000000
	^{13}C	1.10	13.003355
Hydrogen	^{1}H	99.98	1.007825
Deuterium	^{2}H	0.01	2.014102
Oxygen	^{16}O	99.76	15.994915
	^{17}O	0.04	16.999131
	^{18}O	0.20	17.999160
Nitrogen	^{14}N	99.63	14.003074
	^{15}N	0.37	15.000109
Sulfur	^{32}S	95.02	31.972071
	^{33}S	0.75	32.971458
	^{34}S	4.21	33.967867
	^{36}S	0.02	35.967081
Phosphorus	^{31}P	100.00	30.973762
Silicon	^{28}Si	92.23	27.976927
	^{29}Si	4.67	28.976495
	^{30}Si	3.10	29.973771
Fluorine	^{19}F	100.00	18.998404
Chlorine	^{35}Cl	75.77	34.968853
	^{37}Cl	24.23	36.965903
Bromine	^{79}Br	50.69	78.918338
	^{81}Br	49.31	80.916292
Iodine	^{127}I	100.00	126.904470

of possible molecular formulas. High resolution analyzers therefore are often used in organic synthesis, where they can verify that the molecular formula of the compound conforms to what was expected.[6–8]

High resolution is also used for the precise quantification of compounds that are members of a vast congener family and generate mass spectra with very complex isotopic distributions, e.g., dioxins, furans, and polychlorinated biphenyls (Figure 4.1).[9] As an example, the quantification of the dioxin referenced as 2,3,7,8-TCDD (tetrachlorodioxydiphenyl) and whose structure is presented in Figure 4.2, is precise only in high resolution because this molecule, recognized as the most toxic of the 75 dioxin congeners, is difficult to separate from its tetrachlorinated congeners and from other chlorinated compounds. Only high resolution allows the precise isolation of the ion used for quantification of 2,3,7,8-TCDD among the overlapping of complex isotopic distributions.

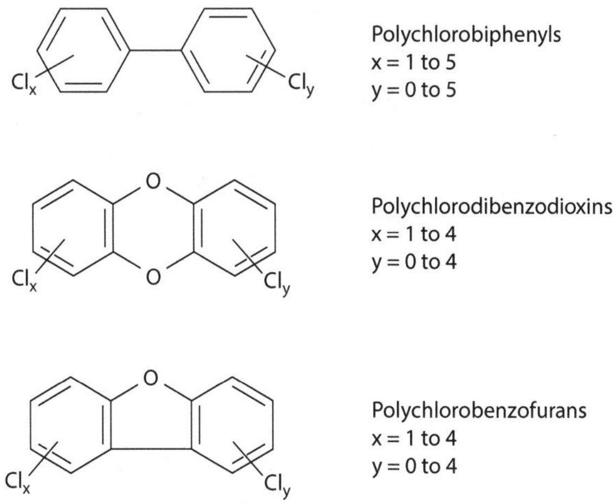

Polychlorobiphenyls
x = 1 to 5
y = 0 to 5

Polychlorodibenzodioxins
x = 1 to 4
y = 0 to 4

Polychlorobenzofurans
x = 1 to 4
y = 0 to 4

FIGURE 4.1 Chemical structures of polychlorobiphenyls, dioxins, and furans.

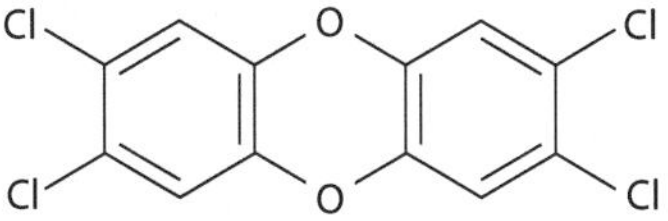

FIGURE 4.2 Chemical structure of 2,3,7,8-TCDD dioxin.

Under the same principle, high resolution will also help detect a specific compound within a very complex mixture (essential oils, extracts of petroleum, wine, coffee) or from an extract containing many matrix interferents, where the probability of meeting co-eluted molecules supplying isobar ions is not negligible. To a certain extent, we can consider that the mass spectrometer will terminate the separation role of the chromatograph.

4.3 QUADRUPOLES

4.3.1 ION TRANSFER OPTICS

As Figure 4.3 shows, a quadrupole is systematically associated with an extraction and focalization system for the ions produced within the source. It is generally a hexapole or an electrostatic lens (electric analog of an optical lens) or a combination of both. This system continuously accelerates the ions produced in the source toward the quadrupole. It is estimated that the average time of residence of the ions within the source is in the range of 10^{-6} second.

The voltage applied to the extraction system determines the polarity of the accelerated ions. A negative voltage is used to extract positive ions in electron ionization and in positive chemical ionization and a positive voltage to accelerate negative ions in negative chemical ionization. An ion beam of the same charge systematically

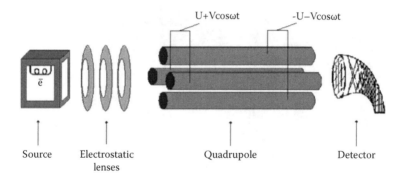

FIGURE 4.3   Quadrupole mass spectrometer.

diverges because the ions repulse each other by Coulomb repulsion. Besides accelerating the ions, the function of electrostatic lenses and hexapoles is to focus the beam in such a way that the ions do not disperse before entering the quadrupole. The extraction system does not exert selection on the ions. All the produced ions access the entrance of the quadrupole; it is only at the entrance of the latter that the ions are separated according to their m/z ratios.[10]

4.3.2   PRINCIPLES OF OPERATION

A quadrupole is composed of four parallel metallic electrodes connected two by two, of hyperbolic or cylindrical section. The hyperbolic section quadrupoles theoretically are the most efficient. These electrodes are from 10 to 20 cm long, depending on the model. Figure 4.4 is a photograph of a quadrupole.

FIGURE 4.4   Photograph of a quadrupole. Scale is shown by comparison with a 1-euro coin.

The application of a potential $\Phi_0 = U + V\cos\omega t$ creates a quadrupolar field between the electrodes. Two electrodes, symmetrical with respect to the central axis of the quadrupole, are brought to the potential Φ_0 and the two other electrodes are brought to the potential $-\Phi_0$. U and $V\cos\omega t$ are, respectively, the components of the DC and AC voltages of the potential.[11] V and ω are, respectively, the amplitude and the pulsation of the AC voltage; $\omega = 2\pi f$. The value of the frequency f is fixed by the manufacturer and is generally around 1.1 MHz (that is why we usually refer to AC as radiofrequency).

The polarity of the electrodes is quickly reversed. This leads the ions on an oscillating trajectory whose radial amplitude depends on the parameters U and V. In order for an ion with a given m/z ratio to follow a stable trajectory in the quadrupole and reach the detector, the U and V parameters must allow the radial trajectory of the ions to be inferior to the distance separating the electrodes. The quadrupole functions like an ion filter and may be referred to as a quadrupole filter.

In practice, one can simultaneously vary the values of U and V, while conserving a constant U/V ratio, in order to keep the ions stable in turn in the analyzer. At a given moment t, only ions of a given m/z ratio are detected. The other ions collide with the electrodes or the internal walls of the manifold of the spectrometer; they discharge and the resulting neutrals are led by the pumping system. A simplified theoretical approach of ion separation is given below.

4.3.3 Theoretical Approach to Ion Separation

Let us consider an ion of a given m/z ratio subject to a quadrupolar field resulting from the application of the potential $\Phi_0 = U + V\cos\omega t$ on the quadrupole electrodes. Two opposite electrodes are separated by a distance $2r_0$ (see Figure 4.5). The ion is subject at the point of coordinates (x,y,z) to the potential $\Phi = \Phi_0 (x^2 - y^2)/r_0^2$.

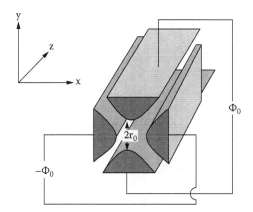

FIGURE 4.5 Quadrupole trunk under sagittal sectioning.

The equations of the ion movement within the field (established from the fundamental relation of the dynamics) in each of the x,y,z directions are the following:

$$m (d^2x/dt^2) + 2ze (U + V\cos\omega t) x/r_0^2 = 0$$

$$m (d^2y/dt^2) - 2ze (U + V\cos\omega t) y/r_0^2 = 0$$

$$m (d^2z/dt^2) = 0$$

where m and z are the mass and the charge of the ion, respectively, t is the time, and e is the electron charge (e = 1.6×10^{-19} Coulomb).

The point here is not to resolve this equation in an algebraic manner. This equation admits two kinds of solutions: (1) those referred to as stable, corresponding to trajectories allowing the ions to go through the quadrupole and reach the detector and (2) unstable solutions corresponding to trajectories allowing the ions to exit the field and thus the quadrupole before reaching the detector. In this last case, the ions are obviously not detected.

When changing variables in order to make the parameters a and q appear, the movement of the ion within the quadrupole field is described by the following Mathieu equations:

$$(d^2x/d\tau^2) + (a + 2q.\cos 2\tau) x = 0$$

$$(d^2y/d\tau^2) - (a + 2q.\cos 2\tau) y = 0$$

with

$$\tau = \omega t/2$$

$$a = -8zeU/mr_0^2 \omega^2$$

$$q = -4zeV/mr_0^2 \omega^2$$

The a and q are called stability parameters; a is directly proportional to U and q is proportional to V. The stability of an ion of a given m/z ratio in the quadrupolar field depends on these two parameters.

We now consider the graphic solutions of the Mathieu equations. Figure 4.6 shows the shapes of the stability diagrams of the ions in the x and y directions (within the orthonormed system shown in Figure 4.5) according to the parameters a and q.

For an ion to pass through the quadrupolar filter, it is necessary for it to be stable in both the x and y directions. The stability zones in the quadrupolar field therefore graphically correspond to the intersection zones of the two diagrams in Figure 4.6. Figure 4.7 shows these intersection zones. We can see that each of the stability diagrams in Figure 4.6 presents three horn-shaped zones: one small, one average size, and a large one. The biggest horns are in the top of the diagram for the stability

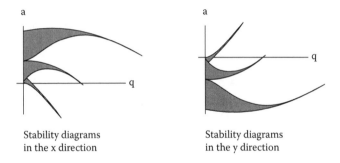

Stability diagrams
in the x direction

Stability diagrams
in the y direction

FIGURE 4.6   Stability diagrams of ions in x and y directions according to parameters a and q.

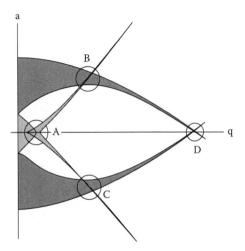

FIGURE 4.7   Intersection zones (A, B, C, D) of stability diagrams of Figure 4.6.

zones in the x direction and at the bottom of the diagram for those in the y direction. Since none of these two zones crosses the q axis, there will be no intersection points between them if we superimpose the two diagrams of Figure 4.6. Consequently, the zones corresponding to the large horns in Figure 4.6 do not appear in Figure 4.7.

Figure 4.7 shows four intersection zones marked A, B, C, and D. Commercial mass spectrometers function under zone A, first because it is the largest and because it corresponds to absolute values for a and q that are inferior to those of zones B and C and to a value of q inferior to that of zone D. From a practical view, working under inferior values of a and q allows use of inferior tensions for the field Φ_0. The stability parameters a and q are directly proportional to U and V.

More specifically, since r_0 and ω are constant, the fact that a given m/z ion does or does not go through the quadrupole to reach the detector depends only on the values of U and V. Consider Figure 4.8 showing the stability diagrams of two ions: m/z 50 and m/z 100 (enlargement of zone A of Figure 4.7). Three working points (A through C) are presented. Working point A (250;700) corresponds to U = 250 V

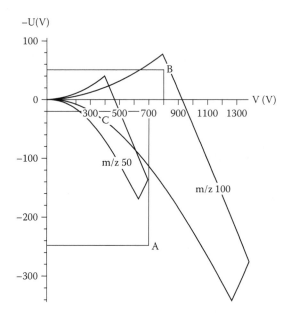

FIGURE 4.8 Stability diagrams of m/z values of 50 and 100 ions.

and V = 700 V; it does not belong to the stability diagrams of the m/z 50 and m/z 100 ions. This means that the application on the electrodes of a field $\Phi_0 = U + V\cos\omega t$ with U = 250 V and V = 700 V (ω fixed at 1.1 MHz) does not allow the detection of m/z 50 or m/z 100 ions.

Working point B(–50;800) is inside the stability diagram of m/z 100. At this working point, the m/z 100 ions are detected and the m/z 50 ones are not.

Working point C(20;400) is inside an intersection zone of the stability diagrams of the m/z 50 and m/z 100 ions. At this point, the m/z 50 and m/z 100 ions go through the quadrupole and reach the detector simultaneously. Since the detector that counts the ions is incapable of differentiating them, it is imperative to avoid working points such as C in order to record an exploitable mass spectrum.

To record a mass spectrum, the quadrupole is subject to a succession of working points that ensure, in turn, the stability of the different ions to be detected. Figure 4.9 represents the superior apices of the stability diagrams of three ions of consecutive m/z ratios: 149, 150, and 151. To review each of the three diagrams, we scan the U and V values while maintaining a constant U/V ratio. This graphically follows a line of the y = ax type.

In Figure 4.9, four straight lines are presented and designated a through d. Line a presents too high a slope: none of the working points situated on line a are in the stability diagrams. Scanning on line a is equivalent to scanning nothing since no ions will reach the detector. Scanning on line d is not satisfactory either. The d line has a U/V slope that is too weak. Many operational points are common to diagrams of ions of different m/z ratios. The scanning line must therefore present a slope within those of lines b and c.

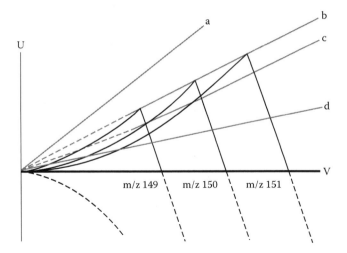

FIGURE 4.9 Superior apices of stability diagrams of three ions with consecutive m/z ratios of 149, 150, and 151.

As this slope gets closer to that of b, more resolution is achieved at the expense of the number of detected ions. In following this line, one spends a lot of time on working points that are outside stability diagrams and scanning no ions.

When the slope of the scanning line is close to that of c, we detect more ions at the expense of resolution. When scanning in this manner, we go through working points that are very close to the intersection zones of stability diagrams. An ion at m/z 150 then risks detection at m/z 149 or at m/z 151. The stability diagrams are theoretical depictions based on a unique ion in an ideal quadrupolar field in an absolute vacuum.

In reality, several thousands of ions repulse each other by Coulomb forces and the field and the vacuum are not ideal. Consequently, the trajectory of an ion in the quadrupole is not rigorously the one defined by the Mathieu equation and it is best to keep a security margin according to the slope of the scanning line to avoid resolution problems.

The parameter that determines this slope is to be fixed by the user in the acquisition method. It is generally named peak width in programs and expressed in m/z. A peak width value of 1.0 m/z signifies that, for instance, ions with m/z values ranging from 149.5 to 150.5 (150.0 ± 0.5) will be detected as m/z 150 ions. In the same manner, a peak width value of 3.0 m/z signifies that the ions of m/z ratios between 148.5 and 151.5 (150.0 ± 1.5) are collected. Increasing the peak width value is called opening the quadrupole and generally exhibits a spectacular gain in the number of detected ions.

In practice, when one tries to develop a quantification method with a detection threshold as low as possible, opening the quadrupole is generally an excellent technique. In this context, the decrease in resolution that accompanies the increase in peak width is not really a problem insofar as the quantification is undergone only on one or two targeted ions.

Surely it would be comforting for a beginner mass spectrometrist to conclude this part dedicated to the theory of ion separation in the quadrupole filter by a reassuring consideration: most quadrupole users have never seen or have long forgotten the Mathieu equations! To program an acquisition method, the user must fix only a few parameters: the start time of the chromatographic recording, the ion range to be scanned, the peak width value, and the mass spectra acquisition frequency (studied in Chapter 5). The program takes care of all the calculations allowing the conversion of these data into working points!

4.3.4 QUADRUPOLE CALIBRATION

Calibrating a quadrupole involves introducing into the source a compound whose ions produced in EI or in CI are known and establishing a correlation between the m/z ratios of these ions and the values of the working points (U;V) at which the ions are detected. This can be achieved by plotting a function of the type (U;V) = f(m/z). During the analysis, determining the values of points (U.V) at which an impact of ions is detected allows accessing through this function to determine the m/z ratios of the considered ions.

Calibration is generally performed automatically. The reference compound for the calibration of GC-MS analyzers is perfluorotributylamine [$N(C_4F_9)_3$)] also known as PFTBA or FC43. This compound is used for four main reasons:

1. Fluorine possessing only one natural isotope, the ion peaks can be identified with no ambiguity by the software that pilots the calibration.
2. Perfluorotributylamine is liquid and is therefore simple to manipulate in standard temperature and pressure conditions and its vapor tension is such that the surface vapor of the compound is easy to introduce in the mass spectrometer via a capillary tube equipped with an electrovalve.
3. Just like all perfluorinated compounds, it presents important anti-adhesive properties that prevent absorption on the source's inner panels.
4. FC43 ionized in EI presents the advantage of fragmenting itself to produced ions of m/z ratios that cover the whole calibration range (m/z 69, 131, 264, 414, 464, 502, and 614).

4.4 ION TRAPS

Two types of ion traps can be used as analyzers in the context of GC-MS coupling. When the ions are produced in a source equivalent to that of a quadrupole before introduction into the trap, we speak of an ion trap with an external source. When the ions are produced directly in the heart of the trap (the compounds are eluted in the trap as they exit the chromatographic column), the trap plays in turn the role of source and analyzer and acts as an internal ionization ion trap.[12,13] Recent ion traps can function in either internal or external mode and also in a hybrid ionization mode whose operation principle is described next.

4.4.1 ION TRAPS WITH INTERNAL IONIZATION

An ion trap is composed of three metallic electrodes: a ring electrode and two endcap electrodes. Spacers in the shape of rings play the role of electric insulators between these electrodes (Figure 4.10).[14] The two endcap electrodes are pierced in their centers—one to introduce the electrons sequentially and the other to allow the ejection of the ions toward the detector. A photograph of an ion trap is presented in Figure 4.11.

Unlike the quadrupole, the ion trap has the ability to store the ions. Applying a radiofrequency of the type $V\cos\omega t$, with $\omega = 2\pi f$ ($f = 1.1$ MHz) on the ring electrode produces a quadrupolar field within which each ion acquires an oscillating motion whose amplitude and frequency depend on the m/z ratio of each ion and the value of V. The more weight of an ion, the more significant is its inertia toward the inversion of the quadrupolar field. Consequently, heavy ions are trapped close to the center of the field, that is, near the center of the ion trap.

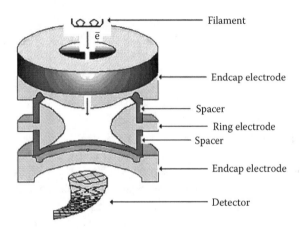

FIGURE 4.10 Representation of an internal ionization ion trap.

FIGURE 4.11 Picture of an ion trap. Scale is shown by comparison with a 1-euro coin.

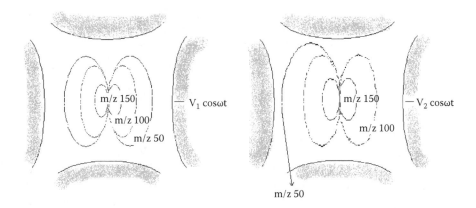

FIGURE 4.12   Trajectories of three ions of different m/z ratios (50, 100, and 150) at V_1 and V_2 values with $V_2 > V_1$.

For a given V value, each ion possesses its own oscillation frequency called secular frequency that depends on the m/z ratio of the ion. The scanning of the value of V leads to an increase in amplitude of the ion's trajectory. Because the ions of weaker m/z ratio present the most external trajectories, they are the first to be ejected. Figure 4.12 illustrates the simplified trajectories of three ions of different m/z ratios and the consequences on these trajectories when increasing the amplitude V of the radiofrequency.

Note that this kind of ion trap does not allow working in electron attachment because thermalizing the electrons while storing them is impossible. Due to its very low mass, a thermalized electron has a stability diagram that is too small for the electron to be trapped and is therefore instantly ejected from the ion trap. Consequently, its time of residence in the trap is very short and the probability that it might meet a molecule it can attach itself to is too low for efficient electron attachment.

4.4.1.1   Cooling Effect of Helium

The quadrupolar field is homogeneous only in the center of the ion trap. Ions should not approach the electrodes or their trajectories will become unstable and they will not be stored. To prevent this situation, one must introduce helium at a partial pressure of around 10^{-3} torr into the analyzer. The ions undergo multiple collisions with helium atoms. The collisions reduce the amplitude of their trajectories and confine them to the center of the analyzer.[15] With ion traps using internal ionization, the carrier gas eluting from the chromatograph also plays the role of cooling gas. In any case, special attention must be taken to ensure that very high purity helium (and no other carrier gas) must be used.

The helium flow rate in the analytical column must be set to a value providing a partial pressure of about 10^{-3} torr in the ion trap (i.e., about 1.0 mL.min^{-1} for a 30-m capillary column with an internal diameter of 0.25 mm and about 1.5 mL.min^{-1} for a 60-m capillary column with the same internal diameter). Finally, calibration of the mass analyzer must be carried out each time the helium flow rate is modified in the chromatographic method. As a matter of fact, changing the partial pressure of

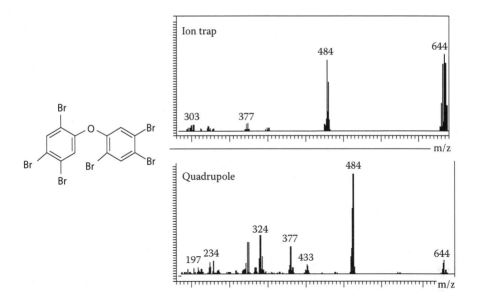

FIGURE 4.13   Electron ionization mass spectra of a hexabromodiphenylether (formula on left) recorded with an internal ionization ion trap and a quadrupole.

helium in the trap reduces (lower flow rate) or increases (higher flow rate) the cooling effect of helium and thus modifies ion trajectories and secular frequencies.

Generally speaking, analyzers of different kinds supply comparable mass spectra for a given molecule. This is fortunate because otherwise a given spectra database (refer to Chapter 8 dedicated to databases) would be useful only with a single type of analyzer which would limit experimentation. In reality, the characteristics of a spectrum depend on the fragmentation rates of molecular ions and therefore on the internal energy supplied to the ions. As seen previously, the source conditions determine the spectrum results. The analyzer functions only to separate the ions. Nevertheless, exceptions to this rule can be observed.

The following example is especially spectacular. Figure 4.13 compares two mass spectra of a hexabromodiphenylether recorded in electron ionization. One was recorded with an ion trap working in internal ionization and the other with a quadrupole. Note that the mass spectra, both resulting from electron ionization, are very different. The one recorded with the quadrupole presents many fragment ions and a non-abundant molecular ion. The other spectrum recorded with an ion trap is dominated by the isotopic distribution of m/z ratios corresponding to the molecular ion and a single fragment ion of notable abundance. This can be explained by the fact that the molecular ion is thermalized by collisions with the helium atoms in the ion trap; it therefore dissipates part of its vibrating energy, which makes it fragment less than under a secondary vacuum.

4.4.1.2   Axial Modulation

During the ejection phase corresponding to the detection of ions, the value V of the trapping potential of the ions is progressively increased while a radiofrequency of

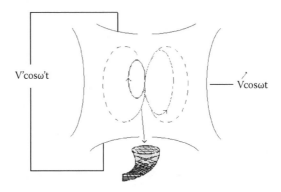

FIGURE 4.14 Principle of operation of axial modulation.

type V′cosω′t (ω′ = 2πf′) is applied between the endcap electrodes. The scanning of V leads all the ions in turn (and of progressive m/z values) to enter in resonance with ω′ when their secular frequency reaches the value of f′. Thereafter, a sudden increase in amplitude of the ion trajectory leads to the ejection of the ions toward the detector (Figure 4.14).

Axial modulation plays a double role. On one hand, it increases sensitivity by ejecting about 50% of the ions toward the detector, whereas bringing the ions at the limit of stability by the simple scanning of V leads them to leave the trapping field in all directions. Sensitivity is also increased due to ion acceleration by resonance that leads them to hit the detector with a higher kinetic energy and thus improves the electronic signal generated. On the other hand, axial modulation improves resolution because resonance ejection is specific to a given frequency and therefore specific to a precise m/z ratio. The value V′ is generally optimized around 4 V for a maximum of resolution and sensitivity.

The most recent ion traps on the market are equipped with a system referred to as triple resonance, which operates differently from axial modulation. Triple resonance allows ejection of all the ions toward the detector (not the 50% with axial modulation), thus doubling the transmission factor.

4.4.1.3 Automated Regulation of Ion Trap Filling

The gap between the electrodes is only a few millimeters so the volume available for trapping of ions is minimal. The first ion traps were not as commercially successful as planned due to charge space phenomena: the electrostatic repulsion exerted by each ion toward the other ions provoked spectral distortions when the analyte was introduced in large amounts. Modern ion traps are equipped with automatic gain regulators (AGCs) that regulate the number of ions present in the analyzer by adjusting the duration of ionization based on the number of molecules entering the trap. Figure 4.15 illustrates the principle of operation of an automatic gain controller.

The first step of operation of AGC is the addition of a "pre-scan" to the traditional recording sequence of a mass spectrum in an ion trap. The duration of the pre-scan must be short in comparison with the total duration of the sequence to keep the spectrum acquisition time reasonable. The pre-scan starts with an ionization phase of fixed

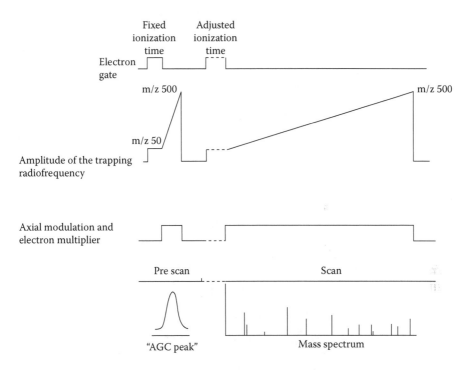

FIGURE 4.15 Principle of operation of automatic gain controller.

duration (100 µs, for example). At the end of ionization, the ions are rapidly ejected on the detector. The ejection speed is much higher than that used in a standard scan because the point here is to count the ions without trying to differentiate them. The objective is then to evaluate quickly the quantity of substance entering the ion trap. This evaluation allows calculation of the ionization time needed to yield a given number of ions referred to as target. The target is fixed by the user in the acquisition method. In an acquisition sequence using AGC, the ionization time that precedes the ion scan is variable. It is automatically adjusted in such a way that the number of trapped ions corresponds to that of the target value.

In the example of Figure 4.16, the target value fixed by the user is 20,000. At a retention time corresponding to Spectrum 1, the pre-scan measures 5000 ions with a fixed ionization time of 100 µs. The algorithm calculates that the ionization time required for scanning is 400 µs (100 × 20,000/5000). At a retention time corresponding to Spectrum 2, the pre-scan measures 30,000 ions; the ionization time is then fixed at 67 µs (100 × 20,000/30,000). At a retention time corresponding to Spectrum 3, the pre-scan measures 200,000 ions; the ionization time is then fixed at 10 µs (100 × 20,000/200,000). At a retention time corresponding to Spectrum 4 recorded in the chromatographic noise, the pre-scan measures 200 ions; the ionization time should then be fixed at 10,000 µs (100 × 20,000/200). In reality, the latter is fixed at 25,000 µs.

This value of 25,000 µs is fixed by the user in the acquisition method as the maximum ionization time. It is indeed important to supply a maximum ionization time to prevent too much acquisition time when no analyte enters the trap.

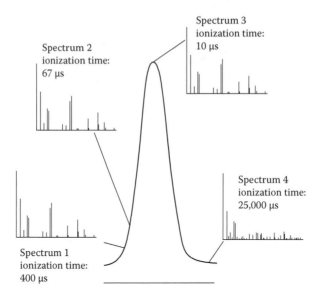

FIGURE 4.16 Example of ionization time regulation by automatic gain controller.

With AGC, the mass spectra are always recorded with the same number of ions, no matter what quantity of analyte is injected. The trace of the chromatograph takes into account the optimized ionization time to correct the value of the measured ion current in order to conserve the quantitative character of the method. Thanks to the electronic gain regulators, ion traps now generate satisfactory spectral reproducibility. However it remains inferior to the reproducibility of quadrupoles, as described in Chapter 6 comparing analytical performances of quadrupolar analyzers.

4.4.2 Theoretical Approach of Ion Separation in an Ion Trap

The theoretical principle of ion separation in an ion trap is close to what has been presented for the separation of ions in a quadrupole. In both cases, a quadrupolar field is used. From a strictly theoretical view, ion trap is nevertheless a simpler technique. On most commercial devices, the quadrupolar field is created by the application of a potential Φ_0 as $\Phi_0 = V\cos\omega t$ (and not $\Phi_0 = U + V\cos\omega t$ as in the case of quadrupoles) on the ring electrode.

Consider the stability parameters such as those described in Section 4.3.3 from the Mathieu equations. In the case of ion traps, a = 0 since U = 0. Only the stability parameter q, directly proportional to V, remains:

$$q = -4zeV/mr_0^2 \omega^2$$

The m and z are, respectively, the mass and charge of the ion, e is the charge of the electron (e = 1.6×10^{-19} Coulomb), ω is the angular frequency ($\omega = 2\pi f$, f is generally 1.1 MHz) and r_0 is the radius of the ring electrode. e, r_0 and ω are constant and the stability of a given m/z ion in the quadrupolar field depends only on the value of V.

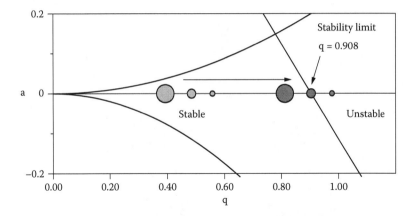

FIGURE 4.17   Stability of three ions as a function of the q parameter in an ion trap.

Figure 4.17 shows how three ions with various m/z ratios (small, medium, and large) leave their stability diagrams when the value of q is increased, which corresponds to an increase of the V value of the radiofrequency. Each ion is ejected when the value of q reaches 0.908. It is to be noted that the y axis corresponding to the parameter a is displayed on the diagram to make the analogy with quadrupolar filters. In fact, scanning is operated only on the x axis corresponding to q values since a = 0.

4.4.3   Ion Traps with External Ionization

Ion traps with an external source appeared on the analytical chemistry market a few years after their internal ionization homologues. This type of device separates ionization and detection, as in the case of a quadrupole. The ions are extracted from the source, accelerated, and focalized by an extraction system analogous to those described for quadrupoles. An "ion gate" permits the sequential introduction of ions into the trap.

The number of ions introduced can be adjusted automatically by an AGC system (see above) that regulates the number of trapped ions not according to ionization time (as with internal ionization traps) but according to the time of opening of the ion gate. Figure 4.18 represents an external ionization trap. As with ion traps using internal ionization, ion storage must be enhanced by a cooling gas. When the capillary column is connected to an external source, helium is introduced into the trap through a line reserved for that purpose.

Compared to their internal ionization homologues, external ionization ion traps present the advantage of being less subject to contamination of the system by non-volatile pollutants that mainly adsorb onto ion volume and ion extraction system. In terms of ionization, the external ionization ion trap does not allow switching from EI to CI in a few seconds but allows electron attachment if one wishes to work in negative mode.

Because the ions are not formed directly in the center of the quadrupolar field, some of them are lost between the source and the analyzer. Some ions fragment themselves on this trajectory because they go from a source under a pressure of

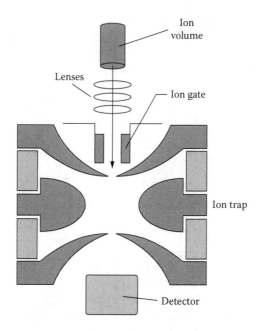

FIGURE 4.18   Representation of ion trap with external ionization.

around 10^{-6} torr to an ion trap containing a partial pressure in helium of 10^{-3} torr (a factor of 1000 between the two pressures). The internal energy gain of the ions during their collisions with the helium atoms at their entrance in the trap is therefore not negligible, especially as the ions must undergo a consequent acceleration to penetrate the quadrupolar field. This translates itself into spectra that present fragmentation rates that are somewhat superior to those of spectra recorded with quadrupoles or internal ionization ion traps. This kind of system generally competes with ion traps using internal ionization when one works with samples that are potentially very pollutant.

4.4.4   Ion Traps with Hybrid Ionization

Recently, ion traps with hybrid ionization have appeared on the market. They offer the possibility of working under internal ionization, external ionization, or a combination of both. The changes in configuration are simple. This kind of ion trap combines the advantages of both systems; it is very efficient in terms of detection limits and very robust because its performances resist the presence of interfering compounds from the matrix.

Ion traps with hybrid sources offer particularly interesting possibilities for chemical ionization. For example, they can perform routine negative chemical ionization which is not electron attachment. Let us consider an example that uses methanol as the reagent. In the source, Reactions (4.1) and (4.2) produce the CH_3O^- anion that can react in the trap with the molecule M according to three reaction pathways: proton abstraction (4.3), charge exchange (4.4), or nucleophilic substitution (4.5). The reaction depends on the nature of the analyte.

$$CH_3OH + \bar{e} \ (70 \ eV) \rightarrow (CH_3OH)^{-\cdot *} \tag{4.1}$$

$$(CH_3OH)^{-\cdot} * + CH_3OH \rightarrow CH_3O^{\cdot} + CH_3O^- + H_2 \tag{4.2}$$

$$CH_3O^- + MH \rightarrow CH_3OH + [M-H]^- \tag{4.3}$$

$$CH_3O^- + MX \rightarrow MX^{-\cdot} + CH_3O^{\cdot} \tag{4.4}$$

$$CH_3O^- + MX \rightarrow CH_3OX + M^- \tag{4.5}$$

One of the applications of proton abstraction is illustrated below. Figure 4.19 shows that a chlorophenol subject to electron attachment gives an ion M^{-} that fragments to supply an ion Cl^- that is absolutely not specific to the analyzed molecule and therefore does not allow characterizing the corresponding chlorophenol. Abstraction of a proton provides an ion $(M - H)^-$ at m/z $M - 1$, which is more stable than M^{-} and especially more specific to the molecule of chlorophenol.

Furthermore, chemical ionization in hybrid mode allows the selection of a reactant ion in the trap, for example CH_5^+ for methane in the positive mode, rather than making the analyte react with each of the ions classically produced by methane (CH_5^+, $C_2H_5^+$, $C_2H_3^+$, and $C_3H_5^+$) when one submits it to electron ionization and allows it to react at a pressure of 1 torr. This is particularly valuable for obtaining a simple CI mass spectrum or carrying out reactivity studies.

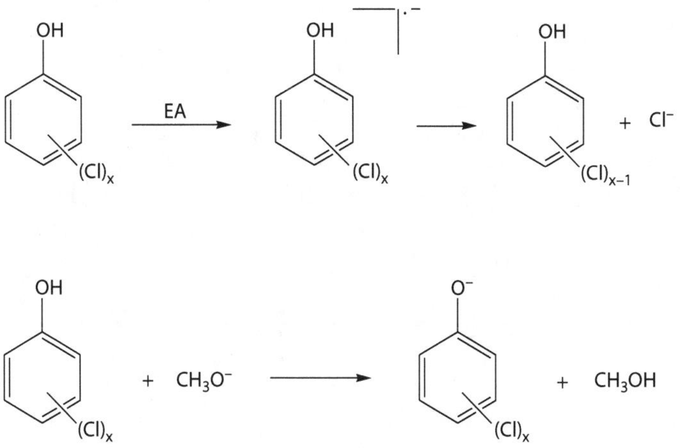

FIGURE 4.19 Ionization of chlorophenol by electron attachment (EA, top) and by proton abstraction (methanol negative chemical ionization, bottom).

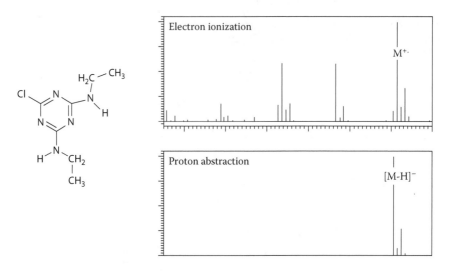

FIGURE 4.20 Mass spectra of simazine. Electron ionization mass spectrum (top) and methanol negative chemical ionization (bottom).

One method even uses the ion m/z 131 of perfluorotributylamine (a compound used to calibrate mass spectrometers) to ionize diclofenac in a hybrid source.[16]

Figure 4.20 compares the electron ionization spectrum and the methanol-negative CI mass spectrum of the simazine herbicide (a monochlorinated compound). The first spectrum is rich in fragment ions, the second only supplies the pseudomolecular ion corresponding to the deprotonated molecule.

4.5   TIME-OF-FLIGHT ANALYZERS

Time-of-flight (TOF) analyzers are more recent additions to the GC-MS coupling market than their quadrupolar homologues. They are now used more frequently in LC-MS, where they are often associated with quadrupoles (Q-TOF), than in GC-MS. GC-TOF systems nevertheless present very interesting applications and are becoming less costly. There is no doubt that TOF and Q-TOF analyzers have bright futures due to the performances they offer for the analysis of complex mixtures (see Section 4.5.3) and structural investigations. We will present the principle of TOF analysis—a far simpler principle than the one that operates quadrupole analyzers—and describe its main advantages and limitations.

4.5.1   ION SEPARATION

A TOF analyzer is depicted in Figure 4.21. The ions produced in the source are conducted toward the acceleration zone where they are subject to a voltage V (around several thousands of volts) applied between two gates. The ions acquire a kinetic energy E_{cin} equal to zeV where ze corresponds to the charge of the ion. The ions then pass through the flight tube at the speed acquired during their acceleration. The

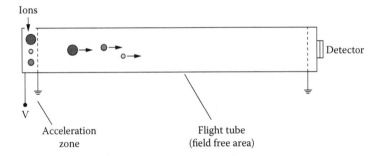

Ions

Detector

V

Acceleration zone

Flight tube (field free area)

FIGURE 4.21 Representation of a TOF analyzer.

flight tube is a field-free area under vacuum. The kinetic energy of an ion is equal to ½ mv² where m and v are, respectively, its mass and speed.

$$½\ mv^2 = zeV \text{ or } v = (2zeV/m)^{1/2}$$

The ion charge and the acceleration voltage are constant. The speed of an ion is as high as its mass is weak; the lighter ions arrive first at the detector situated at the extremity of the flight tube.[17]

L is the length of the flight tube. L varies according to the model of the considered mass spectrometer; it is generally several dozens of centimeters long. The speed v of an ion corresponds to the length L divided by the time t that the ion takes to transit the tube length:

$$v = L/t$$

Consequently, the following relation exists between the m/z ratio of an ion and the TOF:

$$m/z = 2eV/(L/t)^2$$

We can therefore see that one can access the m/z ratio of an ion directly from the measure of its TOF after calibration of the analyzer.

A TOF analyzer functions like a very precise chronometer. The acceleration of ions is pulsated; this means that the voltage V is not applied continuously. It is applied in short pulses. At each pulse, the chronometer is reset to zero and one measures the times of flights of the different accelerated ions when they collide with the detector.

Although TOF analyzers were invented many years ago, the recent devices are much more efficient than the early ones due to the spectacular progress in the fields of electronics and computers. Indeed, the faster the internal clock of a computer, the more precise are the measurements and the better the resolution is.

In a flight tube, the ion beam is much larger than in quadrupole analyzers due to the large distance to be covered. The beam increases naturally during the flight due to the Coulomb repulsions among the ions. Consequently, mass spectrometers that use TOF analyzers are equipped with detectors composed of microchannel plates

(Section 2.3.2) because these detectors are the only types offering large collection surfaces for ions.

4.5.2 ELECTROSTATIC MIRRORS

In the configuration presented previously, the TOF analyzer presents the inconvenience of low resolution. Due to turbulence in the acceleration zone, collisions with neutrals, Coulomb repulsions, and other factors, ions of the same m/z ratios acquire slightly different kinetic energies and reach different speeds and times of flight. This translates to an increase in dispersion of times of flight and a decrease in resolution. Several attempts have been made to solve this problem. "Delayed extraction" achieved some success but the electrostatic mirror surpasses the other inventions.[18]

There are different kinds of electrostatic mirrors, also known as electrostatic reflectors or reflectrons, but they all operate under the same general principle. The mirror is composed of metallic grids on which a voltage gradient is applied. This voltage gradient opposes itself to the kinetic energy of the ions. The ions penetrate the mirror and are restrained. They stop when their potential energy opposes itself exactly to their kinetic energy, then turn back and fly toward the detector. With an axial mirror, the detector can be placed close to the acceleration zone (the ions go back and forth in the flight tube). With an off-axis mirror, the detector is placed at the extremity of a second flight tube. The two configurations are shown in Figure 4.22.

The process of refocusing of ions in kinetic energy can be explained by considering two ions (A and B) of the same m/z ratio. During acceleration of the ions, A acquires slightly more energy than B and flies faster. If the detection is done without a mirror, A will be detected before B. According to the resulting mass spectrum, A and B will not have exactly the same m/z ratios. If an electrostatic mirror is used, A has more kinetic energy than B and thus penetrates deeper into the mirror before

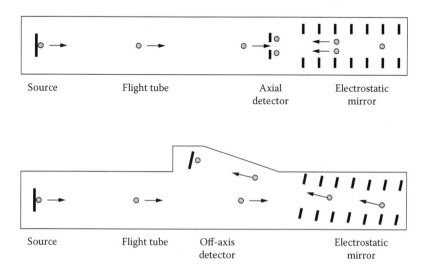

FIGURE 4.22   Electrostatic mirrors. Axial mirror (top) and off-axis mirror (bottom).

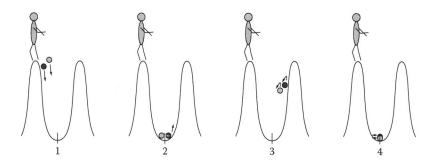

FIGURE 4.23   Mechanics of energy refocusing exerted by electrostatic mirror.

turning back. Finally, the ion that has the most kinetic energy (the fastest) must undergo a longer trajectory in order to compensate. The ions A and B collide with the detector at the same time; they are detected at the same TOF and are therefore given the same m/z ratio.

To explain the principle of ion energy refocusing simply, we can make an analogy with mechanics. Consider a person at the top of a hill holding a marble in each hand (Figure 4.23). The two marbles are identical. When the person lets go of the marbles simultaneously, the black marble moves a fraction ahead of the other one (because one hand is slightly lower than the other or an involuntary push was given). If one detects the passing of the marbles when they reach the bottom of the hill (diagram 2), the faster of the two will logically arrive first. Diagram 4 shows the marbles at the bottom of the hill. If the marbles ascend the side of a second hill and go back down another way (diagram 3), they will be detected simultaneously because the one moving faster had more energy and climbed slightly higher on the side of the second hill. As with ions in an electrostatic mirror, the fastest marble had to undergo a longer trajectory and finally was caught by the other one at the precise moment of their detection.

4.5.3   Advantages and Limitations

By using an electrostatic mirror, the kinetic energy refocusing exerted by the mirror allows high resolution as long as the ions are not scanned too fast. With efficient calibration, it is indeed possible to measure the m/z ion ratios to a precision level of 4 or 5 decimals. This makes the TOF fitted with an electrostatic mirror one of the least expensive, high resolution mass spectrometers on the market.

In GC-MS coupling, the TOF analyzers are not used as much for high resolution as for their fast m/z ratio scanning. When a modern quadrupole records mass spectra at frequencies around 10 per second (for a range of scanned m/z ratios of 50 to 500 Th), the TOF analyzers record mass spectra at frequencies close to 50 per second in many applications and this value can also reach 250 spectra per second! This makes TOF a very efficient technique when a chromatograph displays many coelutions. This is generally the case when mixtures as complex as essential oils and petroleum products (the main uses of TOF analyzers in GC-MS) are analyzed. Along with the performances of deconvolution algorithms of the software with which TOF mass spectrometers are equipped, the high acquisition frequency of spectra yields chromatographic peaks with satisfactory profiles (thin and exhibiting Gaussian peaks) for various complex mixtures.

Thanks to this ability to generate high frequency spectra, the TOF analyzers are particularly useful for coupling with two recent chromatographic techniques: fast GC and two-dimensional GC. Both techniques are generally reserved for the analysis of very complex samples (see Chapter 2).[19]

As a high resolution analogue of the triple quadrupole, the Q-TOF analyzer allows high resolution MS/MS (refer to Chapter 5 covering acquisition modes). A collision cell is placed between the quadrupole and the TOF analyzer. An ion is isolated in the quadrupole (SIM) and fragmented in the collision cell. The daughter ions are analyzed by time of flight, which allows accessing their exact masses. The interpretation of the mass spectrum is therefore considerably simplified and the specificity of the characterization of the analyte is greatly increased.

In spite of the major assets of high resolution and the ability to record mass spectra at high frequencies, TOF analyzers now represent a minority on the GC-MS coupling market. Aside from financial and other considerations, TOF analyzers have a reputation of low efficiency for quantification (quantification of analytes is indispensable for most processes in industrial analysis laboratories), in particular because of the use of microchannel plates to detect ions. Researchers often criticize multichannel plate detectors because they do not supply an electric current proportional to the number of ions that collide with them due to the plate regeneration time. Regeneration involves rehomogenization of the electronic distribution in the microchannels after a massive impact of ions.

Quantification is nevertheless possible with this type of analyzer, as long as one works on weak dynamic ranges (see Chapter 7). Most users of TOF spectrometers in GC-MS coupling speak of semi-quantification. This infers that the spectrometer allows the estimation of analyte quantities rather than a precise dosage. This issue is contested based on certain applications in which TOF analyzers quantify as accurately as quadrupoles. Considering the economic stakes accompanying the efficient counting of ions, it is very probable that detection solutions will soon address this problem.

4.6 MAGNETIC ANALYZERS

Magnetic analyzers were the first types used in mass spectrometry. Their principle of operation is more complex than that of the TOF analyzers that have competed for several years with magnetic analyzers on the high resolution market. It is not possible to make an exhaustive presentation of these analyzers because of the diversity of combinations of magnetic and electric devices. Some are exclusively dedicated to the measure of m/z ratios in high resolution; others are conceived to perform MS/MS (see Chapter 5 dedicated to acquisition modes) in high resolution. One would have to dedicate an entire book to these analyzers to present them in detail and discuss all their applications. This section is limited to a general presentation of magnetic analyzers. We encourage users of such devices to consult books and articles dedicated to the subject.[20,21]

4.6.1   ION SEPARATION IN A MAGNETIC FIELD

Ions accelerated from a source by a tension V acquire a kinetic energy given by:

$$E_{cin} = \tfrac{1}{2}\, mv^2 = zV \qquad (4.6)$$

where m, z, and v are, respectively, the mass, charge, and speed of the ion. The ions enter a magnet oriented in such a way that the magnetic field B exerted by the magnet is perpendicular to the trajectory of the ions (Figure 4.24). Under the effect of the magnetic field, each ion endures a force F_B oriented perpendicularly to its trajectory and to the direction of B. The direction is governed by the "right-hand grip rule." In an absolute value, the force is worth zvB. Under the effect of this force, each ion acquires a circular trajectory of radius r_B. On this circular trajectory, the ions are further subject to a centripetal force Fc, whose amplitude is given by:

$$Fc = mv^2/r_B \qquad (4.7)$$

In order for an ion to go through the magnetic sector, the two forces must balance each other, conditioned by the equality:

$$zvB = mv^2/r_B$$

leading to:

$$r_B = mv/zB \qquad (4.8)$$

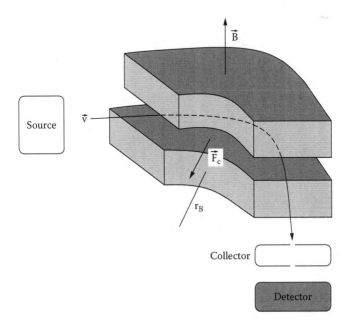

FIGURE 4.24   Magnetic sector.

The radius r_B of the trajectory of an ion in the magnet is therefore proportional to its mass momentum mv, divided by the charge. By squaring the members of the equality and substituting v^2 by 2zV/m after Equation (4.6), we obtain:

$$m/z = r_B^2 B^2/2V \qquad (4.9)$$

The relation (4.9) establishes that the radius r_B of the trajectory within the magnet of an ion of a given m/z ratio depends on B and V. For given B and V values, ions of different m/z ratios will have circular trajectories of different radii. An ion collector composed of a very thin slot is placed at the exit of the magnetic sector. Only the ions exiting within the axis of the slot are detected. By scanning the value of B (most frequent case) or that of V, one detects in turn ions of different m/z ratios.

The calibration of the analyzer from the ions of a known compound allows the measure of the m/z ratios of the detected ions. As mentioned earlier, the ions accelerated from the source always present a certain distribution in kinetic energy at their arrival in the analyzer after collisions or Coulomb repulsions. Due to this distribution in energy, the focalization on the collector's slot is not perfect with ions of the same m/z ratio, which means low resolution. It is thus necessary to combine an electrostatic sector to the magnetic sector to achieve high resolution with this kind of system. The double-focusing mass spectrometer is described below.

4.6.2   DOUBLE-FOCUSING MASS SPECTROMETERS

4.6.2.1   Electric Sector

A double-focusing mass spectrometer combines a magnetic sector such as the one described above with an electrostatic sector. The system is composed of two curved parallel electrodes (Figure 4.25) between which a difference of potential is applied in order for an electrostatic force of a value of E to exert itself perpendicularly to the surfaces of the electrodes. In absolute values, the intensity of this force is worth zE. For an ion to cross the electrostatic sector, this force must match its centripetal force mv^2/r_E where r_E is the radius of the circular trajectory of the ion in the electrostatic sector. This condition utilizes the equality:

$$zE = mv^2/r_E \qquad (4.10)$$

Substituting mv^2 for 2zV [Relation (4.6) linking kinetic energy to the acceleration voltage of the ions] yields:

$$r_E = 2V/E \qquad (4.11)$$

Relation (4.11) shows that, in the case of the electrostatic sector, the radius of the ion trajectory is independent of its mass and charge; it depends only on the V/E ratio. The value of r_E is constant because the position of the collector slot is fixed; the scanning of E allows the selection of ions according to their kinetic energy.

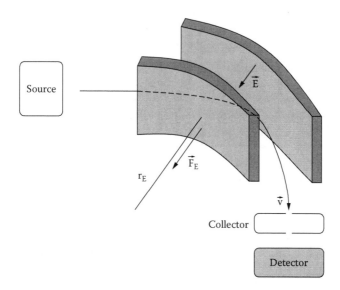

FIGURE 4.25 Electrostatic sector.

4.6.2.2 Principle of Double Focusing

We speak of a double-focusing analyzer because reaching high resolution requires focalization of ions in kinetic energy and direction.[22] Unlike the analyzers described previously, the direction is fundamental to a double-focusing analyzer because it participates in the ion separation. The collection of ions is performed by a very thin slot (thinness necessary to obtain good resolution). The focalization of the ion beam is absolutely necessary. The focalization in kinetic energy is ensured by the electrostatic sector and the focalization in direction is ensured by the magnetic sector.

The configuration of the mass spectrometer presented in Figure 4.26 is one of the most classic double-focusing configurations to reach high resolution, but a wide range of sector devices is available.

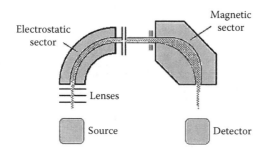

FIGURE 4.26 Scheme of double-focusing mass spectrometer.

4.6.3 ADVANTAGES AND MAIN APPLICATIONS OF MAGNETIC MASS SPECTROMETERS

The main asset of magnetic and electrostatic sectors devices is clearly the high resolution based on double focusing. These devices supply precision measurements of m/z ratios to at least five decimals and thus allow determination of the elementary compositions of ions. The applications of these devices are those described in Section 4.2.4.2 dedicated to high resolution and include the follow-up of organic synthesis and the quantification of organic pollutants like dioxins that have many congeners in biological and environmental matrices.

Where budgets allow, basic research laboratories equip themselves with devices consisting of three or four sectors with collision cells distributed among the sectors. These mass spectrometers can perform high resolution mass spectrometry in tandem (see Chapter 5 on acquisition modes). This means their precision is at least four decimals in the m/z ratio of either the precursor ion and/or the daughter ions produced by collisions.[23]

REFERENCES

1. Cole, R. B. 1997. *Electrospray Ionization Mass Spectrometry: Fundamentals, Instrumentation, and Applications*. New York: Wiley-Interscience.
2. Marshall, A. G., C. L. Hendrickson, and G. S. Jackson. 1998. Fourier transform ion cyclotron mass spectrometry: a primer. *Mass Spectrom. Rev.* 17: 1–35.
3. Marshall, A. G. 2002. Fourier transform ion cyclotron resonance detection: principles and experimental configurations. *Int. J. Mass Spectrom.* 215: 59–75.
4. Miller, P. E. and M. B. Denton. 1986. The quadrupole mass filter: basic operating concepts. *J. Chem. Educ.* 63: 617–622.
5. Chemcalc. 2011. http://www.chemcalc.org/web.
6. Lee, M. L., G. P. Prado, J. B. Howard et al. 1977. Source identification of urban airborne polycyclic aromatic hydrocarbons by gas chromatographic mass spectrometry and high resolution mass spectrometry. *Biol. Mass Spectrom.* 4: 182–186.
7. Beynon J. H. and A. E. Williams. 1963. *Mass and Abundance Tables for Use in Mass Spectrometry*. New York: Elsevier.
8. Gross, J. H. 2011. *Mass Spectrometry: A Textbook,* 2nd ed. Berlin: Springer.
9. March, R. E., M. Splendore, E. J. Reiner et al. 2000. A comparison of three mass spectrometric methods for the determination of dioxins/furans. *Int. J. Mass Spectrom.* 194: 235–246.
10. Dawson, P. H. 1976. *Quadrupole Mass Spectrometry and Its Applications*. Amsterdam: Elsevier.
11. Beynon, J. H. 1960. *Mass Spectrometry and Its Applications to Organic Chemistry*. Amsterdam: Elsevier.
12. Todd, J. F. J. and A. D. Penman. 1991. The recent evolution of the quadrupole ion trap mass spectrometer: an overview. *Int. J. Mass Spectrom. Ion Proc.* 106: 1–20.
13. March, R. E. 2000. Quadrupole ion trap mass spectrometry: a view at the turn of the century. *Int. J. Mass Spectrom.* 200: 285–312.
14. March, R. E. and J. F. J. Todd. 1995. *Practical Aspects of Ion Trap Mass Spectrometry,* Vol. 2. Boca Raton, FL: CRC Press.
15. March, R. E. 1997. An introduction to quadrupole ion trap mass spectrometry. *J. Mass Spectrom.* 32: 351–369.

16. Agüera, A., M. Mezcua, F. Mocholí et al. 2006. Application of gas chromatography-hybrid chemical ionization mass spectrometry to the analysis of diclofenac in wastewater samples. *J. Chromatogr. A* 1133: 287–292.

17. Mamyrin, B. A. 2001. Time-of-flight mass spectrometry (concepts, achievements, and prospects). *Int. J. Mass Spectrom.* 206: 251–266.

18. Karataev, V. I., B. A. Mamyrin, and D. V. Shmikk. 1972. New method for focusing ion bunches in time-of-flight mass spectrometers. *Sov. Phys. Techn. Phys.* 16: 1177–1179.

19. Dallüge, J., R. J. J. Vreuls, J. Beens et al. 2002. Optimization and characterization of comprehensive two-dimensional gas chromatography with time-of-flight mass spectrometric detection (GC×GC–TOF MS). *J. Sep. Sci.* 25: 201–214.

20. De Hoffman, E. 2005. *Kirk-Othmer Encyclopedia of Chemical Technology,* 5th ed. New York: John Wiley & Sons.

21. Cottrell, J. S. and R. J. Greathead. 1986. Extending the mass range of a sector mass spectrometer. *Mass Spectrom. Rev.* 5: 215–247.

22. Burgoyne, T. W. and G. M. Hieftje. 1996. An introduction to ion optics for the mass spectrograph (abstract). *Mass Spectrom. Rev.* 15: 241–259.

23. Boyd, R. K. 1994. Linked-scan techniques for MS/MS using tandem-in-space instruments. *Mass Spectrom. Rev.* 13: 359–410.

5 Acquisition Modes in GC-MS

5.1 DEFINITIONS

5.1.1 INTRODUCTION

After defining the "three S rule" and presenting the concept of signal-to-noise ratio, this chapter describes the main acquisition modes of ions in GC-MS coupling: scanning, selected ion recording (SIR), selected ion monitoring (SIM), selected ion storage (SIS), tandem mass spectrometry (MS/MS), multiple reaction monitoring (MRM), selected reaction monitoring (SRM), and multistage mass spectroscopy (MS^n). These concepts are common to most analyzers and their definitions are valid whether one works with a triple quadrupole or a time-of-flight (TOF) analyzer, for example.

Their application can, on the other hand, differ from one device to another and it is impossible to describe all the possibilities offered by every kind of device in this volume. The more technical aspects of each acquisition mode are detailed only for the systems that fall into the category of quadrupolar analyzers and represent more than 95% of the total GC-MS systems in the world.

5.1.2 THREE S RULE

In order to underline the strengths and weaknesses of the different acquisition protocols, it is necessary to define the three main characteristics of dosage in mass spectrometry. These definitions are also useful for Chapter 7, which discusses the quantification and development strategies of analytical methods. The three S rule is a mnemonic technique for remembering three critical characteristics that start with the letter S: specificity, selectivity, and sensitivity.

Specificity is the reliability of a method in terms of recognition of the molecule to be dosed. The more specific a method, the smaller the risk of dosing an incorrect molecule—a false positive. By extension, we can include in specificity the ability of a method *not* to fail in detecting a molecule that is present in a significant quantity—a false negative. Generally speaking, specificity refers to the characterization of an analyte.

Selectivity is defined as the ability of the dosage method to render results that are independent from interfering compounds, no matter what the nature of such compounds (co-eluted products, matrix interferents, molecules resulting from bleeding of the stationary phase of the capillary column, and others). Note that the selectivity of an analytical method depends on both the parameters relative to GC-MS coupling and the sample preparation process.

Confusion often arises about limit of detection (LOD) and **sensitivity**. In method validation, sensitivity refers to the slope of the calibration line used for quantification. In common language, when we say that an analytical method is *sensitive*, we generally mean that it detects very weak concentrations. In reality, the sensitivity and LOD values are not necessarily in correlation. This explains how dosage method A can be more sensitive than method B when B supplies a limit of detection inferior to that of A (see Chapter 6 dedicated to the comparison of quadrupolar analyzers). The LOD is the smallest concentration at which one can detect an analyte; it is much more interesting than the sensitivity from an analytical view. The lower the limit of detection, the more efficiently a method will detect traces.

It goes without saying that a method of quantification is considered as efficient as it is specific and selective and its LOD is low.

5.1.3 SIGNAL-TO-NOISE RATIO

The signal-to-noise ratio (SNR) is fundamental in method development. There is indeed no point in trying to increase a signal by increasing the ionization current if the noise in the chromatogram increases proportionally. The important task is to obtain a chromatographic peak that is easy to integrate, even at low concentrations. Not all spectrometrists measure the SNR in the same way. As Figure 5.1 demonstrates, there are at least two ways of considering the signal and two ways of considering the noise.

In Figure 5.1, S corresponds to the height measured between the top of the chromatographic peak and the abscissa line that corresponds to no detected ion (no ionic current). S′ indicates the height between the top of the peak and the average level of the base line of the chromatographic trace in the area of the considered peak. N is the height measured between the average level of the base line and the abscissa axis. N′ denotes the amplitude of the oscillation of the base line. S′ and N′ correspond to the signal and the noise as generally defined with an oscilloscope. The noise corresponds to the electronic noise produced by the measuring system itself.

In GC-MS, choosing one kind of measure over another is difficult because under the average oscillation level of the base line, the detected current corresponds primarily to ions that should be considered when determining the SNR. To be rigorous,

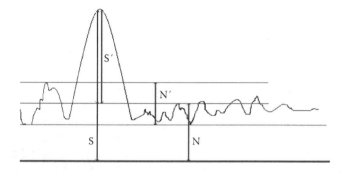

FIGURE 5.1   Two ways of considering the signal (S and S′) and two ways of considering the noise (N and N′) for a chromatographic peak.

we should be able to distinguish the offset, electronic component of the noise, and the chemical noise corresponding to the detection of ions (ions of the residual atmosphere, of the bleeding of the stationary phase, etc.). We should also be able to establish the amount of noise supplied by metastability phenomena in a quadrupole, for instance.

A few portions of (metastable) ions dissociate in the analyzer. The resulting lighter daughter ions are ejected from the quadrupole. However, the neutrals thus generated continue more or less along the initial trajectory of the parent ion. Some of them may reach the detector. The impacts then recorded are correlated to the U and V values applied to the electrodes of the quadrupole and converted in a peak in the mass spectrum. In reality, these impacts are not correlated to the arrival of ions.

The different GC-MS programs therefore do not measure the SNR in the same way. This concept becomes problematic only when one wants to compare the performances of different mass spectrometers on a specific analysis. During method development, it is only important for the user to determine this ratio in the same way at each optimization step of the method to establish the influences of all parameters on the SNR. Logically, an increase in the SNR will generally translate as a decrease in the LOD values of the method.

5.2   FULL SCAN ANALYSIS

5.2.1   DEFINITION OF FULL SCAN RECORDING

The scanning or full scan mode is used for recording source spectra—spectra in which all or most of the ions produced at a given moment in the source are present. The *scanning* term is used because obtaining such a spectrum requires the scan of the voltages U and V applied to the electrodes of the quadrupole, the voltage V applied to the ring electrode of the ion trap, and the value of field B of a magnetic analyzer (refer to Chapter 4). Working in full scan means recording the ions over a large range of m/z ratios. This generally starts at 45 or 50 Th due to the ions resulting from the ionization of the residual atmosphere in the source: m/z 18 (H_2O^+), m/z 28 ($N_2^{+\cdot}$ and CO^+), m/z 32 ($O_2^{+\cdot}$), and m/z 44 ($CO_2^{+\cdot}$) and covers several dozen or hundreds of Thomson values depending on the analytes of interest.

5.2.2   TOTAL IONIC CURRENT

What is a chromatogram recorded in GC-MS? As Figure 5.2 demonstrates, it is a kind of three-dimensional representation of all the mass spectra recorded as a function of time. The abscissa values of the chromatogram correspond to the retention times as in classic chromatography. The ordinate values correspond to current values, the current in each point resulting from the sum of the currents corresponding to each ion of the mass spectrum recorded at that moment.

TIC indicates total ionic current. RIC, meaning reconstituted ionic current, is also used. The third dimension, perpendicular to the plan of the paper, corresponds to the mass spectrum's trace (ion current as a function of m/z ratio). For practical reasons, the chromatograms are never visualized as in Figure 5.2. They are visualized in

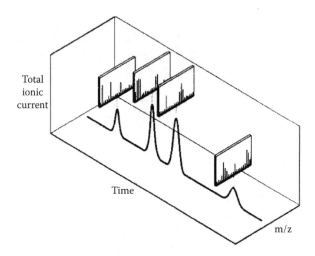

FIGURE 5.2 Three-dimensional representation of chromatogram recorded in GC-MS coupling.

two dimensions: total ion current as a function of retention time. With most GC-MS data treatment programs, clicking on a point of the chromatogram opens a window in which one can visualize the mass spectrum recorded at the retention time corresponding to the point.

5.2.3 Scan Frequency for Data Acquisition

The chromatographic trace resulting from a succession of mass spectra, as seen previously, brings up a question. During the programming of an acquisition mode, at what frequency is it wise to record mass spectra? One scan every 5 seconds? Ten scans per second? The answer is less simple than it seems at first glance. To answer this question, we must first define the scan and microscan parameters in GC-MS programs.

The scan is the time interval between two consecutive mass spectra in a chromatogram. The acquisition frequency (or sampling rate) is generally expressed as scans per second. The microscan is the effective duration of the scan of an analyzer. Its duration depends on the range of m/z ratios to be scanned set by the user for the acquisition method. If, for example, a quadrupole scans at a speed of 6000 Th/second and the acquisition method scans m/z ratios in a range from 50 to 249 Th (200 Th), the duration of a microscan will be 200/6000 = 1/30 of a second.

The microscan is the shortest duration necessary for recording a mass spectrum. The fastest spectrum acquisition frequency corresponds to 1 scan = 1 microscan. By definition, the duration of a scan cannot be inferior to that of the microscan. Therefore, if one programs, for example, a scan of 0.2 second (or a spectrum every 0.2 second) with an ion trap presenting a microscan duration of 0.3 second on the specified scanning range (50 to 500 Th, for example), the software will send an error message because the demanded frequency is superior to that authorized by the performance of the device.

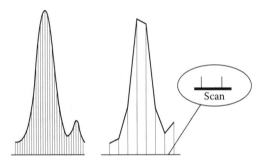

FIGURE 5.3 Chromatogram extract plotted with frequencies of four scans/second (left) and one scan/second (right).

Consider the chromatogram extract presented in Figure 5.3. At a frequency of 4 scans/second, the interval between two points of the chromatogram is 0.25 second; there are 30 points for describing the chromatographic window of interest. In this example, the peak displays a Gaussian shape. The apex (top) of the peak is precisely located as is the lower point between the major peak and the following one that is slightly co-eluted. Let us now consider the same extract of the chromatogram plotted with a frequency of 1 scan/second with an interval between the points of 1 second. There are only 8 points to describe the chromatographic window of interest. The main peak is not Gaussian; its apex and coelution degree (and thus its retention time) are not well defined. Furthermore, it is impossible to integrate the peak accurately.

This example shows that a too-low sampling rate will impair the chromatographic trace. Based on this result, we are logically tempted to record the mass spectra at the highest frequency possible, that is, 1 scan = 1 microscan. This option is not always the right one as the spectral repeatability along the chromatographic peak may be poor due to turbulence, repulsion, and collision problems mentioned earlier. Under the 1 scan = 1 microscan condition, each spectrum results from a simple scan and is thus composed of few ions; it weakly represents the ions contained in the source. When 1 scan = 4 microscans, for example, the recording of each mass spectrum covers four complete scans of the analyzer, which contributes to reducing the disparities between spectra and increases spectral repeatability.

One must also take into consideration that the TIC associated with the mass spectrum under these conditions will be approximately four times superior to that from 1 scan = 1 microscan and that the resulting chromatographic response will be more intense. This last point is obviously very important for trace analysis.

In conclusion, the question of acquisition frequency of mass spectra is fundamental; it is too often pushed aside by novice users who generally program methods using the default sampling rate value of the software. This error often leads to difficulties during the development or validation of dosage methods. The acquisition frequency of spectra will indeed determine the quality of the chromatographic trace and that of the mass spectra. This parameter is directly in correlation with the scanning speed of the analyzer that therefore deserves consideration when purchasing a mass spectrometer.

5.2.4 Purposes

Source spectra are used to identify analytes. Most GC-MS users utilize databases that classify many mass spectra (up to several hundreds of thousands) and allow the instant identification of the analyzed molecules (as long as they are recorded in the given database). If the studied molecules were not previously classified, their identification must be carried out by deduction from observed ions. This is often tedious and requires specific training in spectrum interpretation. The MS/MS techniques described below assist in the interpretation of source spectra. They are often essential to elucidate the structure of an analyte.

The identified compounds can be quantified by GS-MS. In this case, the integration of the chromatographic peak occurs on the current corresponding to an ion that is characteristic of the analyte. As an example, Figure 5.4b demonstrates from the chromatogram of Figure 5.4a how one can observe selectively the peak corresponding to the molecule of clotiazepam by viewing only the current associated with the detection of an ion characteristic of this molecule (m/z 319).

Even coeluted products can thus be precisely quantified as long as their mass spectra differ by at least one ion. It is important not to confuse the action of extracting the chromatographic profile corresponding to the current of a given m/z ion from a chromatogram recorded in full scan, as in Figure 5.4b, with programming the device to record selectively only the current of a single ion (modes SIM, SIR, and SIS described next).

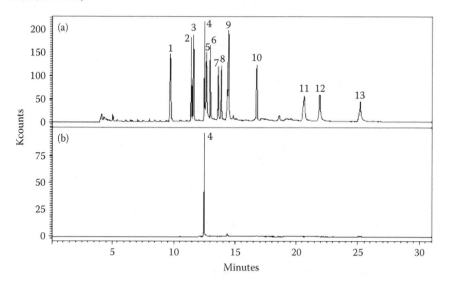

FIGURE 5.4 Chromatograms of a solution of 13 benzodiazepines at 20 μg/mL in acetonitrile. (a) Screening of the benzodiazepines in full scan mode (chemical ionization): medazepam (1), tetrazepam (2), diazepam (3), clotiazepam (4), nitrazepam (5), midazolam (6), flunitrazepam (7), bromazepam (8), prazepam (9), flunitrazepam (10), estazolam (11), alprazolam (12), triazolam (13). (b) Same chromatogram on the extracted current corresponding to the ion at m/z 319 characteristic of clotiazepam.

5.2.5   LIMITATIONS

Let us consider a quadrupole programmed to detect ions by scanning between m/z 50 and 549—a range classically used in GC-MS. The mass spectrum of a molecule M contains 10 ions of m/z ratios between m/z 72 and 289. This choice has no relation to reality because 10 is a weak number in electron ionization and a high one in chemical ionization, but it allows the simplification of calculations.

This spectrum is represented in Figure 5.5. When the quadrupole scans a range of 500 Th, 1/500 of the scanning time is used for the detection of each ion of the range. We can say that during a microscan, the time imparted to the detection of the ions of a given m/z ratio corresponds to 1/500 of the total duration of the microscan.

During the elution time of molecule M, all the ions present in the spectrum displayed in the figure were formed in the source in the same proportions as those observed in the spectrum. All of them are extracted using lenses and enter the quadrupole.

At the moment when the scan passes by the working point corresponding to the value pair $(U;V)_{50}$ associated with the stability of the m/z 50 ions, all the ions issued from the source are ejected from the quadrupolar field because no m/z 50 ions are formed in the source. A moment later, the quadrupole is subject to the working point corresponding to the value pair $(U;V)_{51}$, that is, programmed to let only the ions at m/z 51 go through the quadrupole. Again, all the characteristic ions of M are ejected from the field; no ion is recorded at the detector as M does not produce any m/z 51 ions.

The same concept applies for all working points $(U;V)_i$ up to i = 72. At that precise moment, all the m/z 72 ions that arrive in the analyzer reach the detector. Then the scan reaches the working point $(U;V)_{73}$ and the mass spectrometer detects nothing once again.

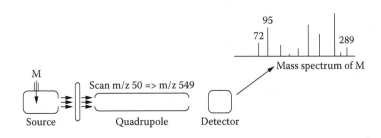

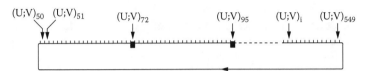

Only 10 working points (U;V) out of 500, from $(U;V)_{50}$ to $(U;V)_{549}$, allow collecting ions resulting from ionization of M on the electron multiplier. During 490 working points out of 500, no ion is detected.

FIGURE 5.5   Limitations of full scan recording with a quadrupole.

We now consider scanning in its entirety. Since M produces only 10 ions in the source, only 10 working points of 500 will allow collection of the ions on the electron multiplier. Consequently, in the chosen example, the quadrupole goes through 490 working points of 500 for which no ion is collected. This means that 98% of the duration of the microscan is dedicated to scanning in a vacuum, i.e., scanning nothing! This example illustrates the Achilles' heel of the quadrupole. This analyzer is not very sensitive for scanning due to its operation under a variable pass bandwidth filter.

Under scanning, the ion trap supplies limits of detection that are significantly inferior because of the capacity of the trap to store ions. Nevertheless, all the molecules eluted in GC-MS with ion traps are not detected because the ionization in the case of internal ion traps, or the refill in ions of the trap in the case of external ion traps, is stopped during detection; the molecules eluted at this time are not detected.

This example illustrates the importance of limiting as much as possible the scanning range in m/z as soon as possible. If we are sure that the analytes of interest have molecular weights inferior to 300, for example, scanning on a range of 45 to 300 m/z rather than 45 to 650 m/z will allow collecting approximately twice the number of ions over the same laps of time. We will see in the following chapter that reducing the acquisition range allows us to increase the ion signal (and thus the size of the chromatographic peak of the analyte) and also reduce the noise and improve the detection limit performance of the analytical method.

5.3   SIR, SIM, AND SIS MODES

The selected ion recording (SIR) mode consists of recording only one or a few selected ions characteristic of the analytes studied. Selected ion monitoring (SIM) and selected ion storage (SIS) are more specific terms commonly employed by quadrupole and ion trap users, respectively.

The gain in sensitivity is spectacular because SIM and SIS increase the signal associated with the detection of analytes while reducing the chromatographic noise. With a quadrupole, the duration of the ion scan is proportional to the range of m/z ratios scanned. Operating on few m/z values therefore increases substantially the time allocated to detecting corresponding ions in comparison with the full scan mode. In parallel, the elimination of the undesirable ions (issued from the stationary phase of the chromatographic column or impurities of the sample) reduces considerably the background noise from undesirable ions and the chromatographic peaks associated with the parasite molecules. It is said that the detection is selective. In an ion trap, SIS reduces the limit of detection because eliminating the parasite ions frees space to store more ions of interest in the analyzer. SIS and SIM are very efficient for trace detection in complex matrices and are also particularly useful for environmental and toxicological analysis.

Let us return to the example used above in the paragraph concerning acquisition limits in full scan operation. If we program a quadrupole in SIM on the m/z 289 ion, for example, the analyzer will be at the working point $(U;V)_{289}$ during the whole duration of the chromatographic segment reserved for the detection of molecule M. Consequently, there will be 500 times more m/z 289 ions detected than when the analyzer is programmed to scan a range of 500 m/z. The signal corresponding to this ion will therefore be about 500 times greater.

In parallel, the noise will be significantly inferior. Indeed, if one considers in first approximation that the global noise results from the sum of the noise components of all the ions susceptible to detection in the m/z scanned range, in SIM there will be left only the "noisy" component of the m/z 289 ion in the chromatogram. This component can be up to 500 times inferior to the global noise. This is only an approximation as all the ions are not involved in the noise in the same way. It is obvious that the contributions to the background noise of the ions issued from the bleeding of the capillary column's stationary phase are vastly superior to those of the other ions.

The gain obtained when one goes from full scan to SIM is quite spectacular; the limit of detection can decrease from a factor of several dozens to several hundreds.

SIM on one ion is the most selective mode; it supplies the lowest limit of detection that can be reached with a quadrupole. It is nevertheless used rarely because it is not very specific. The analyte is characterized only by a single ion and its retention time. It is often insufficient.

Another example involves toxicological analysis. How can we affirm that Mr. X murdered his neighbor under the influence of LSD (lysergic acid diethylamide, a very strong hallucinogen) because we detected in his blood sample a characteristic ion of LSD at its retention time? Human blood is so complex that the probability of detecting a molecule other than LSD that has a retention time close to that of LSD and a common ion is far from negligible. We might as well say that the risk of obtaining a false positive in SIM with one ion is very high in matrix analyses. This lack of specificity is obviously incompatible with toxicological practice and the legal consequences it represents.

This low specificity can be significantly improved by programming the quadrupole to detect several ions selectively. SIM with three ions is widely used. The quadrupole covers three working points in turn. The characterization of the analyte is then more rigorous since it takes into account the chromatographic retention time, the presence of each of the three ions, and also their relative abundances.

In SIS, the ion or ions are detected by applying between the endcap electrodes of the trap a combination of frequencies set to eject all the ions except the one or more selected thanks to a "notch" at the secular frequency or frequencies of the ion or ions to be trapped. Figure 5.6 depicts a notch. Although the isolation of ions in SIM is very precise, the isolation in SIS does not generally render results as good because

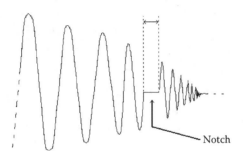

FIGURE 5.6 Notch concept.

the notch is imprecise. The secular frequencies of the trapped ions are very close and it is difficult to eject all the "neighbors" without ejecting the ion itself. A notch covering too large a range of frequencies is not selective. A notch operated on a reduced frequency range is more selective but some ions of interest are lost. The compromise between the two situations is never completely satisfying. Consequently, SIS is generally less efficient than SIM (refer to Chapter 6).

5.3.1   CHOOSING IONS

The choice of ions for SIR is made from a first spectrum obtained in full scan and depends essentially on abundance and specificity. It is obvious that more than one ion is abundant in the source spectrum and choosing it will allow the SIR mode to supply low limits of detection. In the same way, if one wants a method to be specific, it is necessary to choose ions that are specific.

It is generally accepted that an ion possessing a high m/z ratio is more characteristic of the analyte. The M^{+} and MH^{+} ions are systematically chosen in electron ionization and in positive chemical ionization, respectively, as long as they are abundant. One must, of course, avoid ions detected in matrix interferents at close retention times or ions produced by the bleeding of the column's stationary phase. Ions from the stationary phase are easy to find as they dominate the mass spectra recorded in the background noise between the chromatographic peaks.

5.3.2   LIMITATIONS

Generally speaking, a more selective method better supplies low limits of detection because selectivity contributes to reducing the background noise. Every analysis involves a competition between specificity on one hand, and selectivity and the limit of detection on the other. If the full scan mode is very specific (all ions formed in the source are present), it is not very selective and not adapted to the quantification of traces, unlike the SIR, SIM, and SIS modes.

When SIM, SIR, or SIS modes include more ions, the specificity increases at the expense of other characteristics. Indeed, adding an ion does not increase the signal except when the ion is more abundant than those already chosen. If that is the case, why not choose it in priority? Adding an ion increases the noise (indicated by a decrease in SNR) and the risks of interference with ions from matrix compounds. The SIM mode with three ions is commonly used for trace analysis in matrices. It is very efficient for this purpose as long as interfering molecules from the matrix are not too numerous.

The two types of analytical chemistry errors that present serious consequences are false positives and false negatives. With a false positive, an analyte not present in the sample is detected. A false negative occurs when an analyte is not detected although it is present in detectable quantities.

The risk of a false positive is never eliminated in analytical chemistry, but is minimal. To produce a false positive, a molecule not present in the analyte would have to supply three ions identical to those chosen to characterize the analyte at the same retention time as the analyte. The three ions would also have to be present in identical proportions to those found in the spectrum of the analyte. The probability of a false positive under these conditions is very low.

If interfering compounds are abundant and numerous, the risk of a false negative is high. To explain this, we must consider the way the software proceeds to characterize an analyte before the start of dosing.

In a lapse around the retention time of the analyte, the software determines whether a chromatographic peak is present by considering the ionic currents of the three characteristic ions of the analyte (modalities are detailed in Chapter 7). If no peak is present, the analyte is not detected. In the opposite scenario, the ionic currents of all the ions retained for acquisition in SIM are integrated and the areas of the corresponding peaks are compared to the relative abundances of these ions in the reference SIM spectrum. If the relative abundances of the ions correspond, the analyte M is recognized and quantified; otherwise, M is not recognized.

Figure 5.7 illustrates the risk of a false negative in SIM with three ions. This example shows that if an interfering compound B partially co-eluted with molecule A presents a common ion (m/z 200 in the figure) to one of the three ions chosen for the characterization of A in SIM, A will not be identified. We will see below that this risk of false negative can be considerably reduced by using a technique called tandem mass spectrometry.

5.4   TANDEM MASS SPECTROMETRY (MS/MS)

MS/MS considerably reduces the risks of false negative compared to SIR, SIM, and SIS modes. This technique allies the advantages of SIM, SIR, and SIS with those of full scan. It is selective, sensitive, and specific because it supplies under certain conditions mass spectra that are sufficiently rich in ions to characterize an analyte with no ambiguities of any kind. It is consequently the best method for the trace determinations of analytes in matrices.

As an example, Figure 5.8 compares chromatograms of a forensic urinary sample recorded in full scan, SIR, and MS/MS. It clearly illustrates the gain of selectivity obtained in MS/MS in comparison with the other detection modes. The presence of lysergic diethylamide acid (LSD), impossible to detect in full scan and SIR, is clearly highlighted in MS/MS.

MS/MS is also used for structural identification. It is a wonderful tool for determining how different ions of a source spectrum are correlated. Establishing *transitions* means determining the daughters, granddaughters, great granddaughters, etc. of a given ion—establishing a family tree of ions.[1,2]

5.4.1   GENERAL PRINCIPLE

Performing MS/MS consists of isolating an ion (all the ions of a given m/z ratio in fact) in the analyzer and fragmenting it through collisions in an inert gas, atomic gas (helium or argon, generally) or molecular gas (nitrogen). The process is known as collision-induced dissociation (CID).[3] Under the effects of the collisions, part of

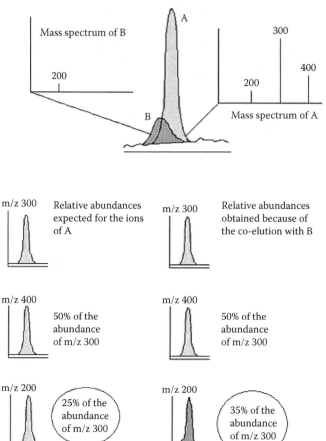

FIGURE 5.7   Risk of false negative in SIM with three ions.

the kinetic energy of the ion is converted into internal energy (see Chapter 2). The increase of internal energy leads the ion to fragment itself. The selected ion is called the parent or precursor ion; it is often a molecular ion but that is not a systematic rule. The ions issued from the fragmentation are called daughter, fragment, or product ions.

The center-of-mass model establishes that the increase of internal energy of the precursor ion depends on its kinetic energy and on the mass of the target atom or molecule of the collision gas.[4,5] Therefore, collisions in argon are much more efficient than those in helium. The nature of the collision gas often depends on the device used. The user must determine a satisfying value of kinetic energy (also known as activation energy) during optimization of the MS/MS method.

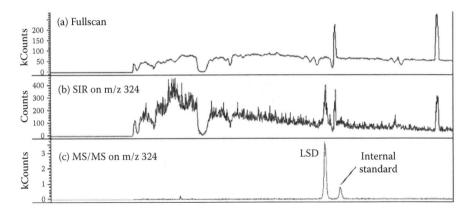

FIGURE 5.8 Chromatograms of a urine extract in positive chemical ionization from a forensic analysis to determine lysergic diethylamide acid (LSD). (a) Chromatogram recorded in full scan. LSD is not detected; the background noise is very abundant. The peaks were not identified. (b) Chromatogram recorded in SIR on m/z 324 (corresponding to MH+ ions of LSD and the LSD isomer internal standard after chemical derivation). (c) Chromatogram recorded in MS/MS activating the ion m/z 324 by collisions.

The MS/MS protocol always entails three steps: (1) the isolation of an ion (precursor ion) that is characteristic of the analyte, (2) the fragmentation of this ion through collisions in an inert gas, and (3) the detection of the obtained fragment ions. The operation of each step differs considerably from one analyzer to another. The three steps are separated in space with a triple quadrupole (one quadrupole per step) and separated over time with an ion trap.

5.4.2 MS/MS WITH A TRIPLE QUADRUPOLE

MS/MS cannot be performed on a single quadrupole. It requires a triple quadrupole (TQ) similar to the one shown in Figure 5.9. The three quadrupoles are generally designated Q1, Q2, and Q3. Q1 is closest to the source and Q3 is closest to the detector. Q2 is also called the collision cell because the activation by collisions takes place in Q2. Although Q1 and Q3 function under a secondary vacuum of 10^{-6} to 10^{-7} torr, Q2 is set to maintain the collision gas at a partial pressure around 1 mtorr. The collision gas can be argon (the most efficient) or nitrogen (less expensive). At this pressure, the system is a "mono-collisional" regime—statistically, most ions are subject to one collision in Q2.

In every case, Q2 works under radiofrequency only (we speak of "rf only"); the potential applied to the electrodes is of type $\Phi_0 = V\cos\omega t$ and not of type $U + V\cos\omega t$ (refer to Section 4.3.3). The consequence is that all the ions are stable in Q2. Indeed, the function of Q2 is not to separate the ions but only to refocus them because their trajectories can often be deviated by the collisions. In certain recent TQ systems, the collision cell is composed of a hexapole to improve focalization of the precursor and daughter ions.

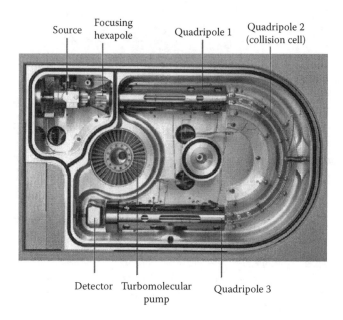

FIGURE 5.9   Triple quadrupole with curved collision cell. (*Source:* Varian Company. With permission.)

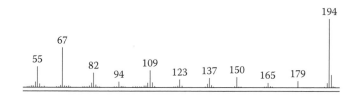

FIGURE 5.10   Electron ionization mass spectrum of molecule M.

The TQ can be used in three modes: (1) to determine the daughters of an ion, (2) to analyze the precursors of an ion, and (3) to scan neutral losses. The operation of each mode is described below. As an example, we will consider the electron ionization spectrum displayed in Figure 5.10 and observe the transitions allowed by each of the operation modes of the TQ operated in MS/MS.

5.4.2.1   Determination of Daughter Ions

The determination of the daughter ions of a selected ion is the operational mode most frequently used with TQs; it is schematized in Figure 5.11.

The first quadrupole is programmed in SIM and selects the precursor ions corresponding to a given m/z ratio. The ions fragment themselves in Q2. Q3 is programmed to scan the m/z ratios of the daughter ions and eventually the non-fragmented precursor ions issued from Q2. The fragmentation yield depends on the internal energy acquired by the precursor ions and therefore on their kinetic energy in Q2. The acceleration of the ions in the collision cell is the main parameter that

FIGURE 5.11   Programming a TQ for determination of daughter ions.

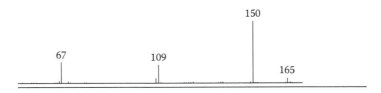

FIGURE 5.12   Mass spectrum of molecule M recorded with scanning of Q3 (Q1 is set in SIM on m/z 165).

must be optimized when developing MS/MS methods. The detector is correlated to Q3. The obtained spectrum displays the daughter ions of the precursor selected by Q1.

If, for example, we select ions of a 165 m/z ratio in Q1 and record the mass spectrum shown in Figure 5.12, this means that ions at m/z 150, m/z 109, and m/z 67 are issued from m/z 165 ions. However, this spectrum does not allow us to determine whether the ions at m/z 67 and m/z 109 are direct daughter ions of m/z 165, are both daughter ions of m/z 150, or if the ion at m/z 67 issued from m/z 109 resulting from dissociation of m/z 150 or m/z 165. All these ions are observed in the source spectrum (Figure 5.10). Performing additional MS/MS experiments on m/z 150 and m/z 109 ions would establish the precise dissociation pathways from m/z 165.

5.4.2.2   Determination of Precursor Ions

The principle of operation of the triple quadrupole to study precursors of an ion is represented in Figure 5.13. The first quadrupole is programmed under scanning. The ions are fragmented in Q2. Q3 is programmed in SIM on the m/z value of the ion whose precursors are researched (m/z 67, for example).

The detector can be hit only by an ion at m/z 67 because Q3 lets no other ions through. An impact with the detector therefore means that a precursor of m/z 67 just passed through Q1. When correlating the impacts on the detector with the U and V values of the potential applied to Q1 at the time of the impact, the recorded spectrum displays the precursors of m/z 67.

FIGURE 5.13   Programming of a TQ for determination of precursors of selected ion.

FIGURE 5.14 Mass spectrum of molecule M, recorded with scanning of Q1 (Q3 is set in SIM on m/z 67).

By selecting the ion at m/z 67 ratio in Q3 while scanning Q1, one records the mass spectrum shown in Figure 5.14. This establishes that the ions at m/z 82 and m/z 165 are precursors of m/z 67. The former example already demonstrated that m/z 67 came from m/z 165; we can see here that some of the ions of m/z 67 issued from m/z 82. The m/z 82 ion is not a daughter of m/z 165 because it does not appear in the spectrum of the fragment ions of m/z 165 in Figure 5.12. It is therefore demonstrated that two dissociation pathways lead to m/z 67 ions and that they are in competition.

5.4.2.3 Neutral Loss Scanning

When we study the loss of neutrals, the quadrupoles Q1 and Q3 are under scanning, as Figure 5.15 displays. The scanning of Q1 and Q3 are separated by the m/z value corresponding to the researched neutral loss. If, for example, we research ions that lose H_2O or are issued from a loss of H_2O, the scanning of Q1 and Q3 are separated by 18 Th. Therefore, when Q1 is at a working point in the stability diagram of the m/z 200 ions, Q3 is at a working point in the stability diagram of the m/z 182 ions. When Q1 filters m/z 201 ions, Q3 filters m/z 183 ions, and so on.

Assume that we are researching the loss of a methyl radical ($CH_3\cdot$) between the ions of the source spectrum in Figure 5.10. Q1 and Q3 are programmed to scan the m/z ratios with a separation of 15 Th. We can correlate the impacts on the detector to the scanning of Q1 and then view the spectrum of the ions that lose a methyl radical. We can also correlate the impacts on the detector to the scanning of Q3 and then visualize the spectrum of the ions issued from the loss of a methyl radical (Figure 5.16). Each of the mass spectra in Figure 5.16 shows that the m/z 67 ion issued from the loss of a methyl radical from m/z 82 and that m/z 150 issued from the same loss from m/z 165.

One may wonder why pursue this kind of experience when it is easy to see the differences of 15 Th between the m/z ratios of these ions during observation of the source spectrum? Because a difference of 15 m/z may be fortuitous and not necessarily result from the elimination of a methyl radical. Therefore, the source spectrum contains ions at m/z 94 and m/z 109 (Figure 5.10). One could think that the

FIGURE 5.15 Programming a TQ for determination of neutral losses.

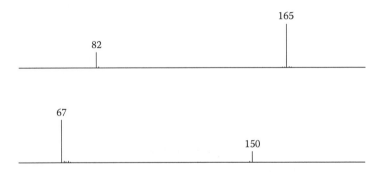

FIGURE 5.16 Mass spectra of molecule M resulting from neutral loss scanning on a triple quadrupole. Losses of 15 amu are recorded. The top spectrum correlates the impacts onto the detector with the scanning of Q1. The bottom spectrum correlates the impacts onto the detector with the scanning of Q3.

formation of the m/z 94 ion results from a methyl loss from m/z 109. The research of neutral losses demonstrates that it is not the case since m/z 109 and m/z 94 do not appear, respectively, in the top and bottom spectra of Figure 5.16.

5.4.2.4 Multiple Reaction or Selected Reaction Monitoring

Multiple reaction monitoring (MRM) and selected reaction monitoring (SRM) are two terms for the same operation. The general goal is to quantify analytes whereas the modes presented below are dedicated to structural analysis. They are not really adapted to quantification analysis because they generally supply limits of detection insufficiently low for trace analysis. We saw at the beginning of this chapter that a quadrupole supplies detection limits that are high when used in the full scan mode due to the principles of operation.

During the development of MRM for quantification with a triple quadrupole, the first step consists of choosing one or two precursor ions, then testing different collision energies to determine which ion supplies the best collision spectrum.

What is the "best" collision spectrum in this context? A spectrum presenting an abundant ion that will be used for quantification and 1, 2, or 3 qualifier ions to be used for characterization of the analyte (see Chapter 7 dedicated to GC-MS quantification). This step is done by programming the TQ in the first of the three modes presented for MS/MS—determining the daughters of an ion. After the choices of collision energy and ions to be detected, we move from MS/MS to MRM by stopping the scanning of Q3 to program it in SIM on the ions of interest, as illustrated in Figure 5.17.

FIGURE 5.17 Principle of MRM or SRM with a triple quadrupole.

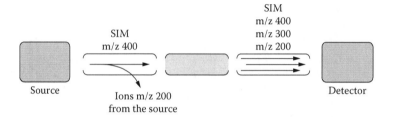

FIGURE 5.18   Programming of the TQ in MRM in the example presented in Figure 5.19.

MRM is very efficient compared to SIM. It considerably reduces the risks of false negatives. If we return to the example in Section 5.3.2 discussing SIM, we can see (Figure 5.7) how even a scarce interfering compound presenting an ion common to one of those retained to characterize and quantify the analyte in SIM can make identification of the analyte fail.

We now consider the same analysis, not in SIM with three ions, but in MRM with collisional activation of the m/z 400 ion and detection of the m/z 400, 300, and 200 ions. Figure 5.18 illustrates the programming of the TQ in MRM in the context of the example. The collision energy is such that the precursor ion is not totally fragmented and thus participates in the characterization of the analyte. The m/z 200 ions issued from the molecule interfering with the analyte do not go through the first quadrupole since it is programmed in SIM on the m/z 400 ratio. Only the m/z 200 ions issued from the fragmentation of m/z 400 ions are detected.

Under these conditions, the relative abundances of the different ions of the analyte are in agreement with those specified in the method despite the coelution phenomenon, as Figure 5.19 demonstrates.

5.4.3   MS/MS with an Ion Trap

5.4.3.1   Collisional Activation in an Ion Trap

Operating in MS/MS with an ion trap involves the addition of an electronic card, not a physical change of the device. While the activation by collisions takes place in a cell when one uses a TQ, activation occurs between the electrodes in an ion trap mass spectrometer. The collision gas must be helium, a gas that is already present in the ion trap as a thermalization component.

This leads to an important question. How can helium be efficient as a collision gas when it was chosen to reduce the kinetic energy of the ions in the trap without fragmenting them? In other words, how can a gas behave simultaneously to induce thermalization and activation? As mentioned in Section 5.4.1, the collisions in helium are not very efficient due to its weak atomic mass. Many collisions at high speed (high kinetic energy) can nevertheless allow a sufficient increase of the internal energy of the precursor ion to allow it to dissociate.[6]

The following metaphor can aid understanding of this concept. Consider a car moving at a speed of 5 km/h on a circular circuit delimited by hay stacks. The car, the circuit, and the hay stacks, respectively, symbolize the ion, the ion trap, and the helium atoms. If the car slides and collides with a hay stack at 5 km/h, the car is

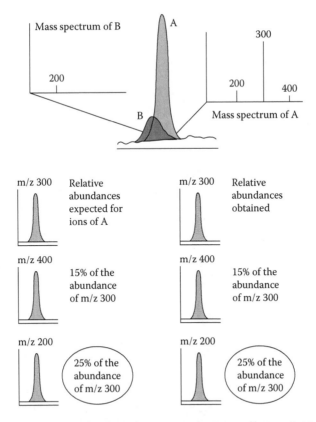

FIGURE 5.19   Illustration of decreased risk of false negative in MRM. Relative abundances of ions of A are in agreement with those expected despite the coelution with B, which contains the m/z 200 ion in common with A. No false negative results.

slowed down but not damaged. At 150 km/h, the same car will collide with dozens of hay stacks before stopping. The intensity of the shocks will deform or destroy the car body. This example shows that a simple change in speed (kinetic energy) can make collisions more "efficient."

To operate MS/MS with an ion trap, we must accelerate the precursor ions sufficiently to allow the collisions with the helium atoms to become activating (and no longer thermalizing).[7] The two ways to accelerate an ion in an ion trap are the resonant and non-resonant modes described below.

5.4.3.2   Precursor Ion Isolation

With an ion trap, the different steps of MS/MS are not separated in space as they are in a TQ; they are separated in time. A typical MS/MS sequence in an ion trap is shown in Figure 5.20.

The first step is isolation of the precursor ion. Isolation is undergone at a certain value of V (amplitude of the radiofrequency applied to the ring electrode). This value corresponds to the m/z ratio below in which the ions are not trapped (m/z 50 in Figure 5.20). After ionization is finished, the value of V is increased in order to

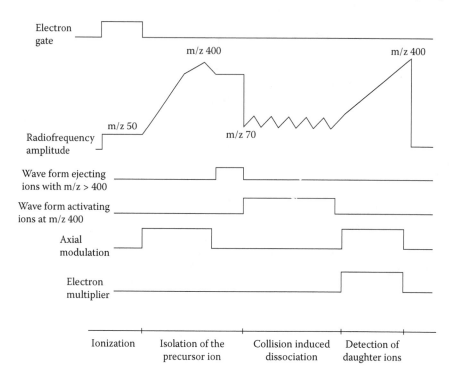

FIGURE 5.20 Sequence of MS/MS with an ion trap mass spectrometer.

eject the ions with m/z ratios inferior to that of the precursor ion (m/z 400 in the example) from the ion trap. The scanning of V slows when the system approaches the value corresponding to m/z 400 in order for the isolation of the precursor ions to be precise.

The ejection of the ions of m/z ratios above 400 results in the application between the endcap electrodes of a combination of frequencies that make all these ions enter in resonance. When the isolation phase of the precursor ions concludes, the amplitude of the radiofrequency of ion trapping is brought back to a value corresponding to m/z 70 (in the example in Figure 5.20). This causes the ions of m/z ratios above m/z 70 to be trapped in the ion trap during the activation phase by collision. The principle of isolation of the precursor ions is the same whether resonant or non-resonant activation is used.

5.4.3.3 Resonant Activation

In the resonant mode, the acceleration of the ions is done by applying a radio-frequency of type $V \cos \omega t$ with $\omega = 2\pi f$ between the endcap electrodes. The value of f corresponds to that of the secular frequency of the precursor ion whose effect is to make it enter into resonance.[9]

Under the effect of the resonance phenomenon, the ion is accelerated and the multiple collisions with helium become activating.[10] The daughter ions issued from the dissociation of this ion have their own secular frequencies that are different from that of

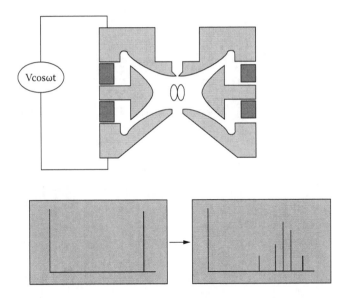

FIGURE 5.21   Principle of resonant activation in ion trap mass spectrometer.

the precursor because their m/z ratios are lower. The daughter ions are therefore not in resonance with the frequency applied between the endcap electrodes. As soon as they are formed, they are thermalized by helium. Consequently, if the activation energy is not too high, the daughter ions do not fragment much, as illustrated in Figure 5.21.

It is to be noted that activation by resonance resembles axial modulation (see Chapter 4) because both require the application of a radiofrequency of type Vcosωt between the endcap electrodes. There are nevertheless two important differences between the two concepts. First, the value of V is around 3 to 4 volts in axial modulation and 0.2 to 1.5 V in MS/MS where the point is *not* to eject the precursor ions. The second difference is the value of the radiofrequency trapping amplitude applied to the ring electrode during the application of the radiofrequency of resonance. This value is high in axial modulation and low in MS/MS where the point is to trap the daughter ions formed by collision.

The resonant mode is generally used for structural analysis because it allows accurate control of the activation energy of the ions. This mode is also used in dosage methods. In this case, the more selective the sample preparation, the more efficient the resonant mode. In the opposite case, the interfering compounds from the matrix present in the ion trap can lead to small differences in the secular frequencies of the precursor ions. This phenomenon is exhibited by a decrease in efficiency of the resonance phenomenon.

The presence of the matrix interferents can diminish fragmentation ratios. To solve this problem, the amplitude of the ion trapping radiofrequency can be slightly modulated during the activation phase by collision. This allows the secular frequency of the precursor ion to be modulated and for the activation to proceed in a range of frequencies instead of at one determined frequency.

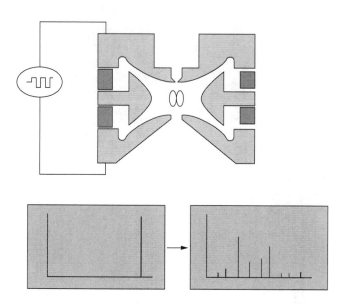

FIGURE 5.22   Principle of non-resonant activation in ion trap mass spectrometer.

5.4.3.4 Non-Resonant Activation

In the non-resonant mode, the acceleration of ions is achieved by applying an electric signal composed of square waves (Figure 5.22) between the endcap electrodes. This has the effect of disrupting the quadrupolar field resulting from the application of the radiofrequency on the ring electrode. The center of the quadrupolar field is quickly moved into the trap. The precursor ions, whose trajectory depends on the position of the center of the field, are quickly accelerated to follow the movement and therefore reach a kinetic energy such that the collisions in helium become activating and fragment themselves.

Unlike the thermalization that operates in resonant mode (see above), the daughter ions are activated after their formation because they are subject to the movement of the quadrupolar field and are accelerated just like the precursor ions. The non-resonant mode therefore encourages the formation of second generation ions (grand-daughters, great granddaughters, etc.) as Figure 5.22 illustrates.

The non-resonant mode is sometimes compared to a shaker. We shake an ion trap violently to fragment the contents. It is generally used for dosage methods because activation by collisions is independent of the secular frequency of the precursor ions. The precursor ions are therefore not affected by the interfering compounds of the matrix eluted in the ion trap.

5.4.3.5 Developing an MS/MS Method with an Ion Trap

Developing MS/MS for structural identification does not require a particular strategy. We isolate the ion whose affiliation we want to establish and dissociate it by collisions with different activation energy values to establish the order of appearance of the fragment ions. The radiofrequency is scanned at a large amplitude to detect all the fragment ions. Each one provides information about the structure of the analyzed molecule.

Developing MS/MS with an ion trap that can be efficient for dosing imposes a few technical requirements. The first development step is the choice of the precursor ion from the source spectrum of the molecule. Obviously we want to choose the most abundant precursor possible after ensuring that it is characteristic of the analyte (and not among the ions that contribute to chromatographic background noise).

The second step is the optimization of CID parameters. Optimization is the same technique whether the activation mode is resonant or non-resonant. This step requires optimization of two parameters: the collision energy and the qz value at which the precursor ions are trapped. The qz corresponds to the component of q according to the z axis of the ion trap (axis through the center of both endcap electrodes (see Chapter 4). This value of qz, often neglected by novice users, is important because it determines the value of the m/z ratio beyond which the daughter ions will be stored in the ion trap. It also conditions the energy that must be supplied to fragment the precursor ion.

Figure 5.23 represents the trajectories of an ion at m/z 400 stored at a qz value of 0.2 and at a qz value of 0.4. At qz = 0.2, the ion oscillates near the center of the ion trap. A "large" space is available around it to trap the daughter ions resulting from the dissociation by CID. The calculations establish that at qz = 0.2 for a precursor ion at m/z 400, all the ions of m/z ratios above 89 will be trapped. At qz = 0.4, the trajectory of the ion at m/z 400 is further from the center of the ion trap and provides less space for the trapping of daughter ions. The calculations show that at qz = 0.4 for a precursor ion at m/z 400, only the ions of m/z ratios above 177 will be trapped.

One must choose the value of qz before optimizing the value of the collision energy because both parameters are tightly linked. Indeed, in resonant or non-resonant mode at a given activation energy value, the fragmentation rate of the precursor ions depends on the value of qz because qz conditions the kinetic energy and thus the initial internal energy of the ions.

The parameters relative to the duration of the isolation and activation phases are also accessible via the software. The default values are generally efficient and it is not necessary to develop an experimental design for the optimization of this parameter to design an efficient MS/MS method.

The third development step concerns the recording of the collision spectrum. It is not possible to work in MRM in the same way we work with a TQ. The detection

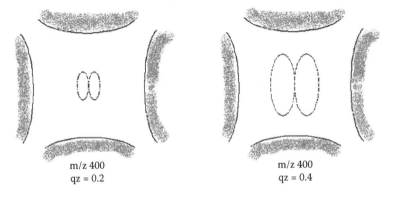

m/z 400
qz = 0.2

m/z 400
qz = 0.4

FIGURE 5.23 Trajectories of m/z 400 ions stored at qz values of 0.2 and 0.4.

of fragment ions implies the scanning of the amplitude V of the trapping radiofrequency. The scanning, as seen previously, is unfavorable for reaching low limits of detection and it extends spectrum acquisition time. Consequently, the aim is to try to reduce as much as possible the range of m/z ratios scanned around the m/z ratios of the chosen ions to quantify and characterize the analyte.

5.4.4 Multiple Stage Mass Spectrometry (MS^n)

The ion trap does not offer the option of studying the precursors of an ion or characterizing neutral losses, unlike the TQ. However, it allows operation in MS^n to reproduce CID experiments in a single operation. For example, an ion of m/z 400 is isolated and fragmented by collisions. Instead of scanning its daughter ions, we will choose one, m/z 250, for example. We isolate it by the ejection of the other fragment ions and fragment it in turn in the same sequence without reintroducing the analyte. This technique is known as MS^3.

This experiment allows in a structural investigation the identification of the granddaughter ions of m/z 400 and the daughters of m/z 250 (m/z 400 can have several daughters). This kind of experiment can also reveal different collision spectra for two molecules (isomers for example) that fragment themselves in the same way in MS/MS.

Figure 5.24 shows the MS^2 and MS^3 mass spectra of a molecule of trimethylsilylated LSD. This weakly volatile molecule was chemically derivatized before

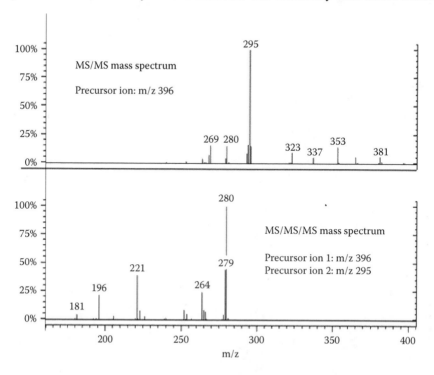

FIGURE 5.24 MS^2 and MS^3 mass spectra (resonant mode) of trimethylsilylated LSD (chemical ionization).

GC-MS analysis. The m/z 396 ion corresponds to the MH$^+$ ion obtained by positive chemical ionization. In MS^3, the m/z 295 ion is isolated and fragmented in turn.

REFERENCES

1. Sleno, L. and D. A. Volmer. 2004. Ion activation methods for tandem mass spectrometry. *J. Mass Spectrom.* 39: 1091–1112.
2. McLafferty, F. and F. Tureček. 1993. *Interpretation of Mass Spectra,* 4th ed. Mill Valley, CA: University Science Books.
3. Bordas-Nagy, J. and K. R. Jennings. 1990. Collision-induced decomposition of ions. *Int. J. Mass Spectrom. Ions Proc.* 100: 105–131.
4. Qian, K., A. Shukla, and J. Futrell. 1990. Collision-induced intramolecular energy transfer and dissociation of acetone molecular ion. *J. Chem. Phys.* 92: 5988–5997.
5. Gilbert, R. G., M. M. Shiel, and P. J. Derrick. 1985. Energy transfer in the collision-induced decomposition of multiatomic ions. *Org. Mass Spectrom.* 20: 430–431.
6. March, R. E. 1992. Ion trap mass spectrometry. *Int. J. Mass Spectrom.* 118: 71–135.
7. Wu, H. F. and J. S. Brodbelt. 1992. Effects of collisional cooling on ion detection in a quadrupole ion trap mass spectrometer. *Int. J. Mass Spectrom. Ions Proc.* 115: 67–81.
8. March, R. E. 2000. Quadrupole ion trap mass spectrometry: a view at the turn of the century. *Int. J. Mass Spectrom.* 200: 285–312.
9. Todd, J. F. J. and A. D. Penman. 1991. The recent evolution of the quadrupole ion trap mass spectrometer: an overview. *Int. J. Mass Spectrom. Ions Proc.* 106: 1–20.
10. Lière, P., R. E. March, T. Blasco et al. 1996. Resonance excitation in a quadrupole ion trap: modification of competing dissociative channel yields. *Int. J. Mass Spectrom. Ion Proc.* 153: 101–117.

6 Comparison of Quadrupolar Analyzer Performances

Criteria for Choosing an Analyzer

6.1 COMPARISON CRITERIA

Although the choice between two high resolution analyzers is generally easy according to the field of activity and the budget dedicated to the purchase, the choice of a quadrupolar analyzer is often difficult for novice users. Choosing between a quadrupole and a triple quadrupole is easy based on the concept that if a triple quadrupole allows MS/MS and MRM, it is possible to make it work as a simple quadrupole if necessary. In this case, the collision cell is under secondary vacuum (no collision gas) and works under radiofrequency only. The first or third quadrupole is also programmed for radiofrequency only; that is, it focalizes ions without operating a selection. The choice between a quadrupole and triple quadrupole is thus easy if the budget allows the purchase of the triple quadrupole!

By contrast, choosing between a quadrupole and an ion trap is far harder in most analytical contexts. The comparison of quadrupolar analyzers in this chapter is based on the criteria of spectral repeatability, response functions, detection thresholds, chromatographic profiles, and maintenance needs.

6.2 SPECTRAL REPEATABILITY AND AUTOPROTONATION

Spectral repeatability is the ability of an analyzer to always supply the same spectrum for a given compound. This repeatability can be estimated by comparing several consecutive mass spectra in the same chromatogram (spectra corresponding to the chromatographic peak of a given compound) or by comparing the spectra of one compound recorded during different experiments.

Spectral repeatability is generally excellent with quadrupoles and mediocre with ion traps, in particular in internal ionization; ion traps favor autoprotonation phenomena. These phenomena generally lead to the observation in a mass spectrum recorded in electron ionization (EI) of an ion at m/z M + 1 corresponding to an ion MH^+ in place or in addition of the $M^{.+}$. The long storage time of ions in an ion trap

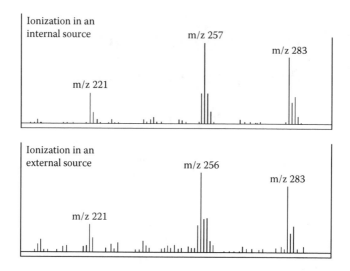

FIGURE 6.1 Comparison of electron ionization mass spectra of diazepam recorded with an internal source (ion trap, top) and with an external source (quadrupole, bottom).

(about 10^5 times longer than the residency times of ions in the volume of an external source) is responsible for autoprotonation. It allows $M^{·+}$ ions to react with M molecules by H· abstraction according to Reaction (6.1). The M molecules are in the majority in the ion trap because of weak ionization yields (see Section 3.2).

$$M^{·+} + M \rightarrow MH^+ + (M\text{-}H)^· \tag{6.1}$$

Autoprotonation can also occur with fragment ions when they possess an odd number of electrons. This case is more rare than autoprotonation of a molecular ion because the daughter ions with odd electron values are generally less numerous than their even number homologues. The dissociation of $M^{·+}$ molecular ions starts more often by cleavages leading to even electron ions (see Chapter 9).[1]

Figure 6.1 compares two electron ionization mass spectra of diazepam, one recorded with a quadrupole, the other with an ion trap operated with internal ionization. The mass spectra are very similar but the base peak is at m/z 256 in external source operation and at m/z 257 in internal source. In an internal ionization ion trap, the daughter ions $F^{·+}$ (m/z 256) react with the molecule M to produce the FH^+ ion (m/z 257). The radical (M-H)· issued from the grasping of a hydrogen radical from the molecule does not, of course, appear because it is electrically neutral.

The formation of protonated molecules in EI can also result from the presence of water (in non-negligible quantities) in the source. The water autoprotonates under EI according to Reaction (6.2) to supply H_3O^+ ions susceptible to transfer a proton to the molecules arriving in the source, depending on Reaction (6.3).

$$H_2O^{·+} + H_2O \rightarrow H_3O^+ + OH· \tag{6.2}$$

$$H_3O^+ + M \rightarrow MH^+ + H_2O \tag{6.3}$$

Autoprotonation phenomena are nonexistent with quadrupoles because the residency times of the ions in the source are very short. They are easy to diagnose with ion traps because the ions resulting from autoprotonation are even more abundant as the molecules of the analyte are abundant in the trap. This is apparent from the fact that these ions are not very abundant in the mass spectra recorded at the bottom of the chromatographic peak (during the ascent or the descent of the peak) and become more abundant as the mass spectra are recorded near the top of the chromatographic peak.

The automatic gain controller (AGC) presented in Chapter 4 regulates the quantity of formed ions. It absolutely does not control the quantity of neutral molecules eluted by the chromatograph and therefore does not allow solution of the autoprotonation problem. Consequently, with an internal ionization ion trap, one must be cautious in accessing a database to find the mass spectrum of a product to be identified from the foot of the chromatographic peak rather than its top, especially if the mass spectra are different according to the position of the peak at which they are observed.

Furthermore, when one develops a dosage method, it is important to ensure that the ion chosen for quantification (see Chapter 7) does not result from an autoprotonation phenomenon. In that case, one should choose another ion or quantify on the ionic current resulting from the sum of radical ion $A^{\cdot+}$ and the AH^+ ion from autoprotonation of the first ion (for example, m/z 256 + m/z 257 for diazepam as described above).

The best way to be free of autoprotonation is to analyze extremely low quantities of matter by injecting weak volumes, diluting the samples, or working under split mode if necessary. Even if autoprotonation puzzles novice internal ionization ion trap users, it is not a problem for experienced users.

6.3   RESPONSE FUNCTIONS AND LIMITS OF DETECTION

The limit of detection of a method (or limit of quantification for a dosage method) is a very important parameter. To compare this parameter in quadrupole and ion trap operations, the acquisition mode of the mass spectrometer must be made precise.

The curve resulting from plotting the chromatographic response of a given compound as a function of its concentration in analyte is called the response function (see Chapter 7). Figure 6.2 compares the response functions obtained with a quadrupole and an ion trap operated in full scan mode for the same molecule.

Figure 6.2 shows a crossing point between the response functions obtained with a quadrupole and an ion trap. The response generated with the quadrupole presents a superior sensitivity according to the technical definition (the slope of the linear part of the response function). This means that at a concentration above the abscissa of the crossing point, the signal obtained with the quadrupole is superior to that from the ion trap. Obtaining a higher chromatographic peak does not present much interest since it can be precisely integrated.

In contrast, Figure 6.2 shows that for concentrations inferior to the abscissa of the crossing point, the ion trap keeps supplying a chromatographic signal while the quadrupole no longer allows detection of the analyte. The limits of detection supplied by the ion trap are therefore inferior to those in a quadrupole under scanning. This results from the ability of the trap to store the ions and accumulate them (thanks to the AGC) when only analyte traces are eluted.

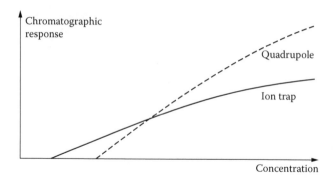

FIGURE 6.2 Response functions obtained in full scan mode with a quadrupole and an ion trap.

In terms of detection limits, the performances of the quadrupole and ion trap are inverted with more specific acquisition modes such as SIM and SIS. SIM in quadrupole supplied limits of detection (and quantification) inferior to those in SIS with an ion trap. The demonstration of performances in SIM is described in Chapter 5 dedicated to acquisition modes. SIS proved less efficient because the isolation of an ion is much less precise with an ion trap than with a quadrupole. Furthermore, the molecules eluted from the chromatograph when the ion trap is not in ionization phase (internal source) or filling mode (external source) are not ionized and therefore not detected. This is not the case with a quadrupole that continuously detects the ions of interest.

In MS/MS, the triple quadrupole allows work in MRM (also called SRM). MRM is the acquisition mode generally used for quantification. It is the most efficient considering the limits of detection and quantification because it avoids the scanning of quadrupoles. It generally supplies limits of detection that are significantly lower than those supplied by ion traps because ion traps cannot liberate themselves from the scanning stage that follows collision and allows ejection of the produced ions onto the detector. The scanning steps reduce the performance in terms of limits of detection, as mentioned above. One must nevertheless be careful in the comparison of the two systems because some exceptions can confirm the rule. A recent study showed that ion traps presented performances in MS/MS that rivalled those achieved with the triple quadrupole in SRM for the dosages of some pesticides in vegetable matrices.[2]

In summary, ion traps supply limits of detection that are inferior to those obtained with quadrupoles in scanning mode. Quadrupoles and triple quadrupoles supply detection limits that are inferior in SIM and MRM, respectively. That is why ion traps are useful for identifying substances and quadrupoles are normally used for quantification. Ion traps can also quantify analytes; quadrupoles and triple quadrupoles can characterize molecules at higher concentration levels.

6.4 CHROMATOGRAPHIC PROFILES

It is generally thought that a chromatogram obtained in GC-MS does not rely on the type of mass spectrometer used because the MS does not participate in the chromatographic separation. We nevertheless saw previously that a mass spectrometer

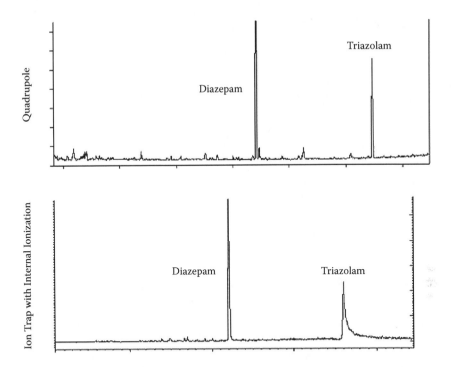

FIGURE 6.3 Chromatograms of mixture of diazepam and triazolam recorded with quadrupole (top) and ion trap with internal ionization (bottom) under identical conditions.

accelerates elution due to the secondary vacuum in the source (see Chapter 2). In reality, a chromatographic profile can be affected by the nature of the analyzer used.[3] Figure 6.3 compares two chromatograms of a mix of two benzodiazepine molecules (diazepam and triazolam) obtained under identical conditions. One of the chromatograms was recorded with a quadrupole, the other with an ion trap with internal ionization.

Figure 6.3 shows an important difference between the two chromatograms. The peak corresponding to triazolam is thin and Gaussian with the quadrupole and tailing with the ion trap. Note that the retention times of the analytes are the same, contrary to what the figure seems to indicate based on different acquisition programs. The tailing of the chromatographic peak of triazolam can be explained by absorption–desorption phenomena in the ion trap.[4]

Indeed, some of the eluted molecules that are not ionized can be adsorbed on the metallic surfaces of electrodes within the ion trap. During the desorption that occurs a few moments later (interactions with the surface are more or less strong), these molecules are ionized. These are the ions formed after the elution of the analyte in the source that are responsible for the tailing of the chromatographic peak.

Experiments have shown that the amplitude of this phenomenon depends essentially on the polarity of the analyte and the temperature. Triazolam is twice as polar

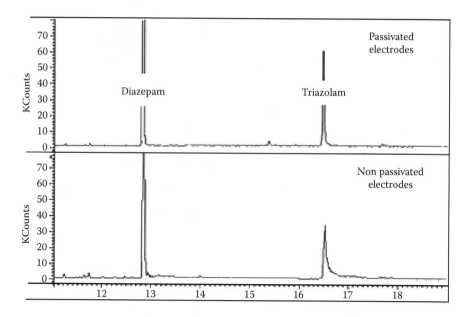

FIGURE 6.4   Chromatograms of mixture of diazepam and triazolam recorded with ion trap mass spectrometer using internal ionization at 220°C. The top chromatogram was recorded with passivated electrodes and the bottom recorded with non-passivated electrodes.

(bipolar moment of 5.8 Debye) than diazepam (bipolar moment of 2.9 Debye). The absorption phenomenon decreases with the increase of temperature in the ion trap since the thermal agitation of molecules limits their adsorption on the electrodes.

Some new generation electrodes are passivated with a silicon-based coating that reduces this type of problem considerably by stopping the adsorption on active sites of metal without eliminating it completely for the most polar molecules. One must nevertheless note that polar molecules are generally chemically derivatized prior to gas phase chromatography analysis (see Chapter 2). Figure 6.4 shows the chromatograms of a mixture of diazepam and triazolam recorded with an internal ionization ion trap at 220°C. One of the chromatograms was recorded with passivated electrodes and presents thin and Gaussian peaks for the two analytes. The other was recorded with stainless steel (non-passivated) electrodes and presents an asymmetrical peak for triazolam.

6.5   MAINTENANCE AND ROBUSTNESS

The pollution soiling of a mass spectrometer is mainly due to adsorption of neutral molecules on the metallic parts. Cleaning these parts is relatively difficult. It generally involves brushing with alumina powder of the ion volume and lens and sonication in organic solvents for the electrodes of ion traps. The electrodes of quadrupoles do not require regular cleaning because the neutrals do not diffuse up to the analyzer. Furthermore, their reassembly after cleaning is a very delicate operation (parallelism must be perfect). The main problem induced by cleaning is the need to stop the pumping system, at least at the level of the source. This means interrupting the analyses for the time necessary, generally a few hours, to allow the mass spectrometer to return under a secondary vacuum.

The spectrometer accumulates more pollution after each analysis, especially if the system is used to analyze matrix extracts. With a quadrupole, the pollution starts at the level of the ion volume before affecting the lenses and the result is an increase in background noise in the mass spectra and therefore in the chromatograms. Sensitivity is therefore reduced but a slightly polluted quadrupole remains operational, especially if operated with an internal standard.

With an internal ionization ion trap, the neutrals adsorb on the electrodes. When the electrodes are polluted, the quadrupolar field is perturbed and the calibration of the analyzer is no longer operational. The mass spectra recorded under these conditions are therefore unusable. In summary, an internal ionization ion trap is not very robust. It does not operate properly even if slightly polluted. For this reason, most applications developed for ion traps concern weakly pollutant matrices, in particular the analysis of water.

6.6 CHOOSING A QUADRUPOLAR ANALYZER

6.6.1 ADVANTAGES

Table 6.1 summarizes the possibilities offered by the main commercial quadrupolar systems. The details of their ionization and acquisition protocols are covered in previous chapters. The chart presents the criteria necessary to help a novice spectrometrist establish rigorous specifications. They do not mention the specific technical characteristics or configurations of specific systems. It is obvious that the best way to form an opinion about a mass spectrometer is to test it (see below).

There are many types of high resolution analyzers with electrostatic and magnetic sectors. It is impossible to cover them all in this chapter because their characteristics vary significantly from model to model and high resolution systems are less standardized than quadrupolar devices.

Ion traps with hybrid sources are not presented in the table. As long as they allow working alternatively in internal ionization and external ionization, they combine the advantages and possibilities of both systems. Rare artifacts can nevertheless complicate the interpretation of mass spectra in hybrid mode. "Ghost" ions can indeed appear in the chemical ionization mass spectra of certain compounds.[5]

Table 6.1 shows that all the devices mentioned allow electron ionization and chemical ionization in positive mode (CI in positive mode is often optional). The internal ionization ion trap offers the double advantage of allowing the use of liquid reagents and rapid switching from one ionization mode to another. This clearly makes the internal ionization ion trap the perfect device for structural elucidation that requires easy access to both ionization modes. However, this analyzer is the only one that does not operate in negative chemical ionization for the reasons covered in Chapter 4.

Concerning acquisition modes, all the analyzers allow analysis under scanning. The quadrupole is less versatile. It allows only SIM whereas the three other systems

TABLE 6.1

**Possibilities Offered by Mass Spectrometers Equipped
with Quadrupolar Analyzers**

	Quadrupole	Triple Quadrupole	Ion Trap with Internal Ionization	Ion Trap with External Ionization
Ionization				
Electron ionization	Yes	Yes	Yes	Yes
Positive chemical ionization	Yes	Yes	Yes	Yes
Negative chemical ionization	Yes	Yes	—	Yes
Rapid switching between EI and CI	—	—	Yes	—
Use of liquid reagents	—	—	Yes	—
Acquisition Modes				
Full scan	Yes	Yes	Yes	Yes
SIM, SIS, SIR	Yes	Yes	Yes	Yes
MS/MS	—	Yes	Yes	Yes
Determination of daughter ions	—	Yes	Yes	Yes
Determination of precursor ions	—	Yes	—	—
Neutral loss scanning	—	Yes	—	—
MRM, SRM	—	Yes	—	—
MS^n	—	—	Yes	Yes

allow analyses in MS/MS (MRM for the TQ) and SIM or SIS. Weakly sensitive in scanning but robust, the quadrupole is thus dedicated to routine dosing.

At a price equivalent to that of a quadrupole, an ion trap offers the possibility of working in MS/MS (sometimes MS^n), is just as practical as the alternatives in terms of ionization, and is more sensitive in scanning. The ion trap also allows alternate SIS and MS/MS operation during the same chromatographic acquisition. That is impossible with a triple quadrupole because the collision cell contains gas in MS/MS and is empty in SIM. The ion trap is much less resistant to pollution and thus less robust than the quadrupole, as noted above.

The choice between ion trap and quadrupole essentially depends on the answers to two questions. First, will one have to analyze polluting samples (e.g., mud extracts, sediments, biological matrices, animal or vegetable materials) routinely? If such materials must be analyzed, the internal source ion trap must be ruled out. The second question is whether interpretation of mass spectra will have to be performed without a commercial database, in which case, the ion trap will be more helpful than the quadrupole. Laboratory specifications are rarely exhaustive and the two following examples show that the choice between quadrupole and ion trap can be difficult.

In the first example, a mineral water distributor realizes after smell and taste tests that a batch of bottles shows a problem. An undesirable substance has apparently contaminated part of the production and possibly the source. What is the device that is best adapted for the identification of the substance responsible for the undesirable taste and smell? The water will have to be analyzed in the full scan mode since the pollutant is not identified. The ion trap will yield detection limits lower than those generated by a quadrupole.

In the second example, the same distributor realizes that a tank of glycol was spilt near the water source. He wants to know whether the glycol polluted the water. What device is the most adapted to answer his question? It is the quadrupole because in SIM it provides lower detection limits than those of an ion trap in SIS. SIM is effective for determining the characteristic ions of glycol.

If one day this distributor decides to equip his laboratory with a GC-MS coupling device to analyze the water, he will face a difficult choice for two reasons. In the first of the two examples, the trap under scanning allows detection of all substances present above a certain limit. The quadrupole allows a significantly lower limit on a substance but only in SIM. In this case, the spectrometer only detects the substance for which it was programmed.

Imagine that the water in the second example was contaminated with a substance other than glycol (e.g., a glycol degradation product). This product will not be detected because the quadrupole in SIM records only the characteristic ions of glycol. By working only under SIM, the risk is not detecting a pollutant even at a high concentration.

The choice between a triple quadrupole and an ion trap depends mainly on considerations relative to MS/MS. It is clear that a quantification method in MRM is more precise than MS/MS methods; furthermore, MS/MS generally supplies inferior limits of detection.

The three operational modes of a triple quadrupole in MS/MS also act in its favor for structural elucidation. The capacity of an ion trap to work under MS^n is rarely exploited by ion trap users. We therefore consider that the triple quadrupole offers more possibilities than the ion trap although the triple quadrupole costs almost twice as much as an ion trap in GC-MS. In mass spectrometry and other analytical fields, economic considerations often play a major role in choices.

6.6.2 TESTING A MASS SPECTROMETER

It is unwise to buy a mass spectrometer before testing it. All mass spectrometer manufacturers welcome potential buyers to try products before they buy them. It is imperative when testing a device to supply appropriate samples, standard solutes, and matrix extracts doped with analytes because results among standard solutions and matrix extracts may differ significantly. It is essential to create conditions as close as possible to those where the equipment will be used. Sample preparation is not the manufacturer's task. Sales representatives are not always equipped for or knowledgeable about sample preparation.

To compare devices of various brands, it is important to make sure that the supplied samples are of the same quality (same concentration, purity, sample preparation protocol). If the specifications require that GC-MS will be used for dosing, one must plan to supply a range of calibration solutions to check the linearity of the response in quantification. This calibration range must be produced from matrix samples doped in analytes and extracted (see Chapter 7 covering GC-MS quantification). If the detection of traces is one of the requirements for purchase, matrix extracts with very low concentrations should be tested.

The test strategies mentioned above are absolutely essential to prevent surprises after purchase. It is wise to request a demonstration of the dismantling and reassembly of the source to ensure that these tasks are not long or difficult. The manufacturer may consider this an unusual request because it requires the device under atmospheric pressure but the ability to dismantle and reassemble a device is very important if it is destined for high flow rate analyses.

The intuitivity of the program interface is another important (and subjective) criterion for choosing equipment. Generally, the part of the software dedicated to settings, use, and parameter adjustments is simple to use. The differences in software performance relate to the level of data processing and quantification along with ease of use.

REFERENCES

1. McLafferty, F. and F. Tureček. 1993. *Interpretation of Mass Spectra,* 4th ed. Mill Valley, CA: University Science Books.
2. Garrido Frenich, A., P. Plaza-Bolaños, and J. L. Martínez Vidal. 2008. Comparison of tandem-in-space and tandem-in-time mass spectrometry in gas chromatography determination of pesticides: application to simple and complex food samples. *J. Chromatogr. A* 1203: 229–238.
3. Borrey, D., E. Meyer, W. Lambert, and A. P. DeLeenheer. 1998. Comparison of quadrupole and (quadrupole) ion-trap mass spectrometers for the analysis of benzodiazepines. *J. Chromatogr. A* 819: 125–131.
4. Libong, D., S. Pirnay, C. Bruneau et al. 2003. Adsorption–desorption effects in ion trap mass spectrometry using in situ ionization. *J. Chromatogr. A* 1010: 123–128.
5. Goulden, P. H., S. Coffinet, Y. Souissi et al. 2011. Investigating the unusual behavior of metolachlor under chemical ionization in a hybrid 3D ion trap mass spectrometer. *Anal. Chem.* 83: 7587–7590.

7 Quantification by GC-MS

7.1 INTRODUCTION

Quantification by GC-MS coupling obeys precise rules. The development of a dosage method by GC-MS can seem long and tedious to novice users, but device programming for quantification does not represent a problem. The preparation of samples generally takes more time and also requires precision and rigor. The validation of the dosage method requires many analyses over several days and can be a source of discouragement, but dosage results can be considered reliable only if the quantification method has been validated first.

Setting up a dosage method includes three important steps : (1) development of the GC-MS method (optimizing parameters for chromatography and mass spectrometry), (2) optimization of the steps of sample preparation (extraction, re-concentration, chemical derivation, etc.), and (3) validation.

As previously noted, sample preparation theory is not presented in depth in this volume. Certain practical issues are discussed in the context of examples. The validation of analytical methods is a very large subject that could fill several books. Method validation is covered in Section 7.4. The discussion is not exhaustive; it presents the basics of validation of a GC-MS method and should allow beginners to learn about validation and determine the quality of their analytical processes.

From a strictly technical view, it would be unreasonable to consider setting up a dosage method in GC-MS without an autosampler. Even if the use of an internal standard makes proceeding by manual injection theoretically possible, practical experience indicates that it is extremely difficult to validate a method through manual injections.

Analytical chemists know that two quantification methods are possible with chromatography: quantification with external calibration and quantification with internal calibration. The first mode is only briefly presented in this chapter because it is not adapted to precise quantification. Priority is therefore given to quantification with internal calibration. The development of dosing methods is discussed elsewhere in this chapter.

7.2 EXTERNAL STANDARD QUANTIFICATION

7.2.1 PRINCIPLE

The principle of external standard quantification is based on injecting solutions of different concentrations and measuring the chromatographic response for each concentration. By plotting the chromatographic response as a function of the concentration

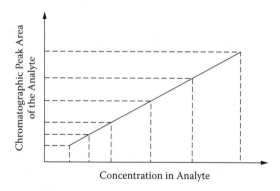

FIGURE 7.1   Building a response function.

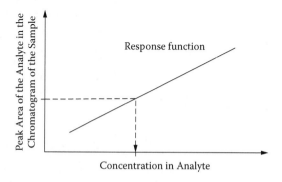

FIGURE 7.2   Determination of analyte concentration in sample using a response function.

injected, we determine the *response function* that is generally a segment of a straight line (Figure 7.1). When performing the analysis to build the response function, it is advisable to inject solutions in an increasing order of concentrations to minimize residual effects.

The question of linearity of the response function will be discussed in the section dedicated to method validation. The concentration of analytes in the sample to be dosed is determined by plotting the chromatographic response of the considered analyte in ordinates onto the response function. The abscissa indicates the researched concentration (Figure 7.2).

7.2.2   LIMITATIONS

External calibration quantification is not advisable in GC-MS because the fluctuations of the chromatographic response may be important. The sample preparation processes generally associated with GC-MS dosage cause many errors that only the use of an internal standard can correct as explained next. In addition to the lack of precision linked to many sources of error with this method, quantification with external calibration imposes a need for regular plotting of the response function (at least daily) requiring numerous analyses over a considerable amount of time.

Analytical chemists familiar with quantification by external calibration will be surprised by these doubts about this method of quantification in GC-MS. It is true that other techniques such as HPLC (high performance liquid chromatography) are more adapted to this calibration mode. Quantification using external calibration seems legitimate in HPLC for several reasons:

1. The injection volume is generally several to several dozens of microliters (versus only 1 μL in GC); the relative uncertainty about the injected volume is lower in HPLC.
2. The discrimination phenomena linked to the change from liquid to gas phase do not exist in HPLC.
3. The ultraviolet and diode array detectors classically used in HPLC offer far more stable responses over time than a mass spectrometer.
4. HPLC methods require less time than GC-MS methods and consequently the repeated analyses required for plotting the response function are less tedious.

Quantifying analyses by internal calibration is desirable whenever possible. Nevertheless, the next section presents the precautions to be adopted when it is necessary to develop a dosage method with external calibration.

7.2.3 SPECIAL PRECAUTIONS FOR EXTERNAL CALIBRATION

We previously explained the necessity of using a standard (whose specificities are described below). In cases where no standards are available, it is necessary to monitor the chromatographic response to ensure its stability during the dosage series. The most rigorous way of proceeding is to trace response functions before and after dosage, and check that the slopes and y-intercepts of the two line segments are equal.

The alternative is to use control samples that have been doped at known concentrations and undergone the same analytical protocol including sample preparation as the samples to be dosed. A first control sample is quantified before the dosage series, a second half way through, and a third at the end of the series. The results of the control sample dosages must obviously correspond to the concentrations expected for the results to be satisfactory. If they are not, the response functions must be built again.

7.3 INTERNAL STANDARD QUANTIFICATION

7.3.1 SOURCES OF UNCERTAINTIES

It would be extremely difficult to determine precisely the global uncertainties in the result of a dosage because of the uncertainties created during all the steps of the analytical method but it is certain that the number would be considerable. As an example, consider a GC-MS dosage of morphine and its main metabolite normorphine (Figure 7.3) in a blood sample. First, the volume of the patient's blood (often 500 μL) to be used for the analysis must be measured as precisely as possible (because the final results of the dosing will be reported to this volume).

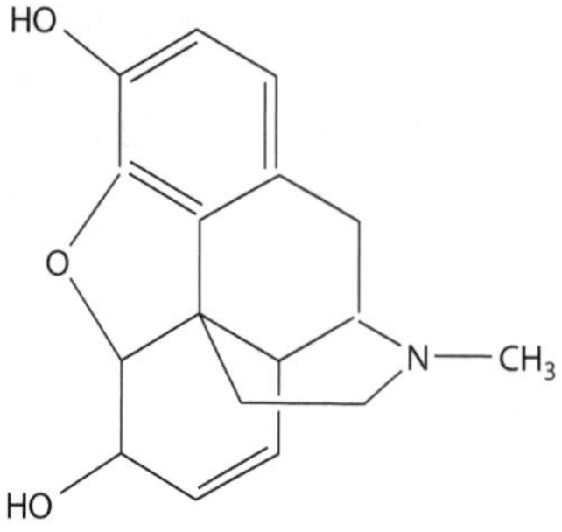

FIGURE 7.3   Chemical structures of morphine (a) and normorphine (b).

The collection of this volume of blood with a pipette induces a first uncertainty based on the precision of the pipette used. Also, collection of blood with a pipette calibrated by the manufacturer for distilled water induces a non-negligible error because the viscosities and densities of water and blood are significantly different.

The second step consists in adding acetonitrile to denature and precipitating the blood proteins. The sample is then shaken and centrifuged for a few minutes. The degraded proteins then form a pellet at the bottom of the tube. The supernatant is collected with a pipette. The uncertainty associated with supernatant collection is important. Two operators will not pipette the supernatant in the same way because the understanding of the limit between the pellet and the liquid is subjective; it is often a pink or brown interface. One operator will never pipette a supernatant the exact same way twice.

The morphine is then extracted from the supernatant by solid phase extraction (SPE).[1] This extraction mode is composed of four steps: (1) conditioning the sorbent of the cartridge, (2) depositing the sample in the cartridge, (3) rinsing the sample with a solvent or mixture of polar solvents to eliminate salts, and (4) eluting the morphine and normorphine with a mixture of organic solvents. Each step introduces uncertainties related to collected volumes and the concentrations of solutions. Crevice creations are frequent in SPE. The sorbent phase may crack, leading to preferential elution trajectories and disruption of the extraction; efficiency can decrease to zero (if the analytes are carried away during the rinsing of the cartridge). In this case, uncertainties led to a false negative error whose consequences can be disastrous during a toxicological analysis!

After SPE, the organic phase containing the analytes is transferred into a small tube and dried under a nitrogen stream at 50°C. During drying, it is possible for tiny quantities of morphine and normorphine to evaporate. The analytes can also be adsorbed on the sides of the glass tube (this phenomenon is common with morphine). The weakly volatile analytes must then be derivatized for analysis by GC-MS. Chemical derivation can be performed with BSTFA [N,O-bis(trimethylsilyl)trifluoroacetamide] in an oven at 80°C (see Chapter 1).

After reaction, the silylation agent is evaporated under a nitrogen stream. Several molecules of analyte may be evaporated or carried away by an azeotropic effect.

A given volume of solvent (generally 30 to 100 µL) is added to dissolve the dry residue for injection in GC-MS. The degree of precision of the pipette used can cause uncertainty during this step.

This example shows that many uncertainties and impacts are associated with the sample preparation process. Another source of uncertainty is error attributable to the autosampler. For example, does the autosampler inject an exact volume of 1.000 µL when it is programmed to inject 1 µL? Probably not.

Injection speed is a critical parameter in GC and a source of uncertainty. Further, injection solvents are generally very volatile and can evaporate partially when a sample is placed in the autosampler, especially when the sample is programmed to be injected several hours after set-up. Errors can also result from an unclean syringe, a partially blocked needle, or a piston that does not slide properly.

Uncertainties can be attributed to the chromatograph, for example, the precision of the temperature of the injector and the oven; the settings of the split valve and electronic flow rate regulator can change over time. Finally, the mass spectrometer induces its own uncertainties in dosage results because the electron current produced by the filament can fluctuate over time, just like fluctuations of the electron multiplier gain. Integration of the chromatographic peak is also subject to non-negligible uncertainties. After a (non-exhaustive!) enumeration of the multiple sources of uncertainty of the dosing results for morphine and normorphine in the example, the need to use an internal standard should be obvious.

Consider the addition of deuterated morphine (morphine-d_3 in which the hydrogen atoms of the methyl group are replaced by deuterium atoms, Figure 7.4) as the internal standard in the test portion of a sample. Most of the uncertainties described earlier will be counterbalanced by internal calibration. An imprecise dilution, a crack in the SPE cartridge, a loss of analytes by evaporation, an incorrect volume injected will all subject a sample to the same losses in standards and analytes. The chromatographic response of the analyte is systematically compared to that of the standard (see next section) so dosing will not be affected by all the uncertainties.

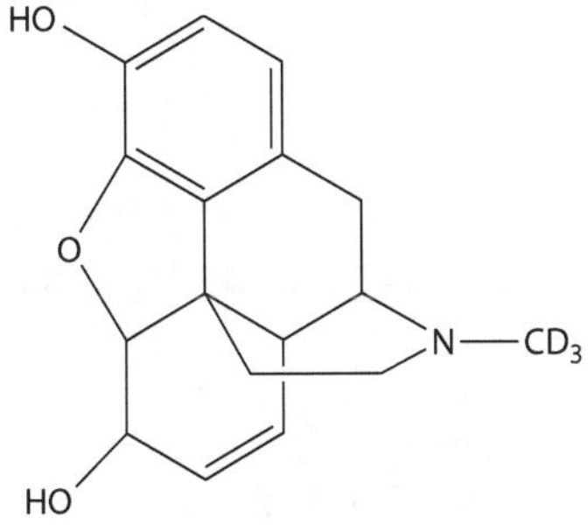

FIGURE 7.4   Chemical structure of morphine-d3.

7.3.2 Principle

As with external calibration, internal calibration relies on the plotting of a response function over a given concentration range, then using the response function to establish the concentrations of the dosed samples in analytes. The fundamental difference between the two dosage methods is based on the fact that in internal calibration, a molecule (standard) is added systematically at a known concentration to the solution to be analyzed.

In internal calibration, we place on the ordinates of the response function the area of the chromatographic peak of the analyte divided by that of the standard peak. It is consequently important for the standard to be added at the same concentration in the calibration solution as in the sample containing the analyte to be dosed.

Using a standard allows suppressing the multiple errors susceptible to falsify the result of the dosing whether they are attributable to the operator or to the device. This also allows the correction of the many errors caused by the sample preparation under the condition that the standard is systematically added at the beginning of the whole analytical process, to the test portion of the sample.

Apart from result precision, the standard fulfills a fundamental function that is often forgotten: confirming the efficacy of the analytical process. Indeed, the detection of the standard demonstrates the absence of major problems such as loss of the analyte during sample preparation, clogged syringe needle, and injection from the wrong vial arising from an error in the programming of the autosampler. The standard must be detected since it was added by the operator. If morphine-d_3 is detected, the absence of a chromatographic peak associated with morphine means that the sample does not contain morphine or contains it at a concentration below the method's detection limit. The failure to detect morphine-d_3 signifies that the analysis failed and that it is thus impossible to detect the presence or absence of morphine in the sample.

7.3.3 Choosing an Internal Standard

It is important for a standard to possess physico-chemical proprieties as close as possible to those of the molecule to be analyzed [polarities, solubilities, sizes (steric effect) for sample preparation, volatilities for gas chromatography, ionization potentials, and basicities for mass spectrometry]. Marketed products are suitable as long as they answer all these conditions.

An important market has developed in parallel to the success of mass spectrometry. Products marked as stable isotopes (usually deuterium or ^{13}C) are sold as internal standards for GC-MS and LC-MS, particularly in the fields of environmental analyses (pesticides, dioxins, hydrocarbons, etc.) and toxicological testing (psychotropic medicines, narcotics, etc.). While these compounds are often expensive, they are used in small quantities and thus the cost may be justified.

Along with their physico-chemical similarities to those of analytes, compounds marked with deuterium or ^{13}C present the advantage that they are not present in significant amounts in the environment and are not endogenous in animals or humans.

This is an important criterion. When developing an analytical method for dosing molecules that have no marked commercial analogues, it makes sense to choose a molecule with a chemical structure close to that of the analyte. It can then be wise to choose a marked product to ensure that the standard is not naturally present in the matrix to be analyzed. If the standard appeared in the matrix at a non-negligible concentration, the dosage results would be false.

Obviously, for practical and economic reasons, one generally does not use as many standards as there are molecules to be dosed, especially if the molecules are present in large quantities. Multi-residue methods that allow simultaneous dosing of 100 molecules or more are used frequently in environmental analysis. Such a method generally uses four to ten internal standards spread out along the chromatogram. Each molecule family is associated with the standard whose chemical structure is the closest to its own.

In the example mentioned above, morphine-d_3 (Figure 7.4) constitutes an ideal internal standard for normorphine and for morphine. Their chemical structures (Figure 7.3) are so close that it is useless to use a standard for each compound.

7.3.4   DEVELOPMENT OF AN INTERNAL STANDARD QUANTIFICATION METHOD

Several development strategies are possible; the one presented below is logical and used frequently. It is adaptable to all kinds of situations and has proven its efficiency according to hundreds of scientific articles.

The example presented in this section concerns the development of a method designed to dose 10 pesticides in lettuce. Six pesticides (numbered 1 to 6) are chlorinated; the four remaining pesticides (numbered 7 to 10) are not. Note that the sequence of the steps does not follow the order used for routine analysis, as illustrated in Figure 7.5. The GC-MS method is indeed indispensable for the evaluation of selectivity and extraction yields of the sample preparation protocol during the development phase.

7.3.4.1   Chemical Derivation

The first question related to the example of 10 pesticides to be dosed is whether they are volatile enough to be analyzed as they are by GC-MS or is chemical derivation required? The answer can be found in the literature. For an experienced chemist, the observation of the chemical structure of a compound can often predict its behavior in GC. The chromatographic peak of the analyte will generally be as thin and Gaussian as it is volatile (low molecular weight and polarity and few or no exchangeable hydrogen atoms). If the answer is not obvious, trials will be necessary to establish whether the compounds are well detected in their current state or choose the best process of chemical derivation if necessary.

Chemical derivation involves methylation, acetylation, silylation, and esterification (refer to Chapter 1 discussing gas chromatography). A derivation agent, catalyst, optimization of reaction temperature (most chemical derivations require heating in an oven), and the kinetics of the reaction must be considered. If two derivation agents supply satisfying chromatography results, one can choose between them according

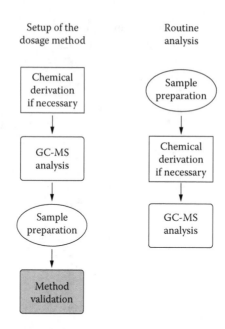

Setup of the   Routine
dosage method   analysis

FIGURE 7.5 Order of steps of method development (left) and routine analysis (right) according to requirements.

to the response in mass spectrometry because the derived products will certainly not have the exact same ionization potential or the same basicity.

It is important to verify that the derivation reaction is complete or almost complete and generates a unique product because a yield of several reaction products considerably complicates the analysis. It increases the number of peaks in the chromatogram and also increases the limit of detection (LOD) by diluting the chromatographic response. Furthermore, with a yield of several products, their relative proportions tend to vary from one analysis to another—a variance that is not compatible with reliable quantification.

When choices of operational conditions are almost complete, it is prudent to proceed with a repeatability test by reproducing the experiment five or six times. All the repeatability tests should be conducted during the validation of the dosage method. If we realize at the end of the validation process that the sample preparation is problematic, we would be obliged to reconsider the whole analytical procedure including sample preparation and the GC-MS development method. The entire validation process would have to be repeated.

If the chemical derivation does not significantly improve the detection of an analyte, we can ignore derivation of the analyte in an effort to gain time and reduce the sources of uncertainties (in the mathematical meaning of the term) of the dosage.[2,3] GC-MS analyses conducted to test the chemical derivation process are generally performed in electron ionization and under scanning on a large range of m/z ratios covering all the possible reaction products to be detected and non-derived molecules if necessary.

7.3.4.2 Developing a GC-MS Method

Developing the GC-MS method requires the preparation of a standard solution containing the pesticides to be dosed and the internal standards. In the example presented, we use two internal standards, one for the chlorinated pesticides and the other for non-chlorinated pesticides. In the standard solution, all the compounds are at the same concentration that is sufficiently high (in the range of ppm, for example) to ensure they will be detected under scanning. If some analytes and standards must be chemically derived, the standard solution should be subjected to the derivation protocol optimized in the previous step.

From this standard solution, one optimizes the chromatographic separation (injection mode, choice of capillary column, and programming of oven temperature), generally in EI because this non-selective ionization mode guarantees the detection of all compounds of interest. Remember, it is not necessary for all the compounds to be perfectly separated since detection operates only on certain characteristic ions. Nevertheless, the better the separation, the less likely the risks of interference. The point therefore is to find a compromise between the best analyte separation possible and an analysis time that is compatible with the specifications of the laboratory.

Errors frequently arise when the heating of the chromatograph oven is stopped as soon as the elution of the analytes is finished. Cutting the heat too soon neglects the fact that future dosages will be operated on matrix extracts that can contain molecules that are less volatile than the analytes. One must plan to elute these molecules, even if they appear to present no interest for the dosage. If they remain adsorbed in the capillary column, they are likely to disturb subsequent analyses. It is prudent to program the temperature of the oven in such a way that, after the elution of the compounds of interest, it reaches quickly (with a gradient of 15 to 20°C/min, for example) the maximal temperature supported by the capillary column (around 300 to 350°C, depending on the column) to evacuate as many matrix interferents as possible.

After the chromatographic conditions are set, the next step is to determine the most efficient ionization mode (EI or CI) for the molecules to be analyzed. The signal-to-noise ratios (SNRs) obtained in EI and in CI for each compound must be compared to the currents of the main ions (not to the total ion current). The better the SNR, the better will be the limit of detection in theory.

At this stage, the choice of ions is frequently a source of error. One type of error is comparing the intensities of the chromatographic signals obtained in EI and in CI. The point is to compare SNRs. Often, the ion current obtained in EI is superior to that obtained in CI but the comparison of the SNRs shows that in reality CI gives better results because it renders far less background noise. Another error arises from comparing the two ionization modes on the total ion currents when the SNRs must be compared on the currents of the ions of interest (ions selected for detection in SIR or precursor ions for detection in MS/MS or MRM). A comparison must be made for each analyte.

In summary, for a method in SIR with three ions, from the chromatograms recorded under scanning in EI and in CI, one extracts two chromatograms for each analyte. The first is reconstituted from the three chosen ions to detect the analyte in

SIR and EI, the second chromatogram from the three ions used for SIR detection in CI. One compares the SNRs of the two corresponding peaks in each chromatogram.

After a comparison is made for each analyte, the next step is to choose the most efficient ionization mode for the method. The choice is easy with an internal ionization ion trap since it allows switching from one ionization mode to another during a chromatographic acquisition. With a quadrupole, the choice can be delicate because certain analytes are detected with lower detection thresholds in EI and some in CI. One must therefore find a compromise to maintain the initial objective of dosing ten pesticides in a single analysis. After the characteristic ions are chosen, we stop working under scanning and adopt the acquisition mode to be used for the dosage.

The next step is making a first estimation of the dynamic range of the method. It involves estimating the detection limit of the method and the range of concentrations on which the response function is linear or almost (linearity of response functions is discussed below) from standard solutions. We prepare a range (perhaps a dozen) of standard solutions of different concentrations. At this step, we can proceed with successive dilutions.

The two internal standards are added after the dilutions because they are always at the same concentration. The different solutions are analyzed by the method chosen by increasing concentrations to avoid residual effects. The limit of detection depends on each analyte; it is estimated from the SNR on the corresponding chromatographic peak. There is no universal convention on the SNR value at which we consider that the limit of detection is reached. This value generally oscillates between 3 and 5, if not 10 for the most demanding analyses. The response function is plotted for each analyte.

These first estimations determine the interval of concentrations at which the sample preparation process must be set up. If sample preparation includes concentration of the analytes, as is often the case, we must consider the concentration factor when doping the matrix (see next section).

7.3.4.3 Sample Preparation

GC-MS is now a routine analysis tool but sample preparation is still the object of many research studies. Sample preparation is generally a complex process of several steps: extraction from the matrix, chemical derivation, filtration, concentration, centrifugation, and so on. The number of sample preparation techniques (and their variations) is too large to detail in this volume.

Among the steps that constitute sample preparation, the only one optimized before the setup of the GC-MS method is chemical derivation (when necessary). This point is obvious since a derivatized compound during ionization does not supply the same ions as it does when not derivatized. The rest of the process must be optimized after the set-up of the GC-MS method.

Extraction often constitutes the key step of sample preparation: it is also the most delicate. No matter what extraction mode is chosen (e.g., liquid–liquid extraction, accelerated solvent extraction, solid phase extraction), it is optimized from a matrix known as a *blank* that is spiked with the molecules to be extracted.

A blank matrix should not contain any analytes that are to be dosed. This condition makes it possible to dope at a precise concentration. The choice of the blank matrix can be delicate for two reasons. The first relates to the uncertainties about the non-contamination of the matrix. It is frequent, for example, to develop a method for the extraction of organic contaminants such as hydrocarbons in river sediment. Real sediment samples are not used because their blank status cannot be ensured. Instead, a sediment is *reconstituted* according to precise norms.

The second difficulty relates to the representativeness of the matrix, especially for biological matrices (blood, urine, hair, viscera, etc.). To develop a method to extract toxins from urine, for example, it is recommended to prepare a blank matrix by pooling a dozen urine samples of various individuals of different ages and genders to obtain an "average" matrix. This is done to accommodate the significant variations of the chemical characteristics of urine from one person to another. To pursue these methodological if not epistemological questions, relative to the reference matrices, refer to a book by Llored titled *Philosophy of Chemistry Practices, Methodologies, and Concepts.*[4]

The concentrations for doping the blank matrix are estimated from the limits of detection (LODs) established by analyses of standard analyte solutions (see Section 7.3.4.2). One must work at concentrations that are obviously higher than the limits of detection able to detect the compounds in standard solutions and consider the concentration factors if necessary. The limits of quantification (LOQs) determined from the matrix extracts during the method validation (see below) risk exceeding those estimated from standard solutions because matrix interferents affect detection. The doping concentrations are also established on the basis of the specifications of the laboratory. It is useless to work at concentrations at which the probability of detecting analytes is nil.

The efficiency of a sample preparation protocol can be estimated on the basis of three parameters: selectivity, extraction yield. and repeatability. During the comparison of protocols, other issues to be considered are the duration of sample preparation, cost, number of samples that can be treated simultaneously, possible automation of steps, natures and quantities of the solvents used, and the environmental consequences.

Selectivity can be estimated subjectively without specific equations. The point is to verify that the chromatographic peaks attributable to the matrix interferents are not confused with those of the analytes on the currents of the ions retained for the characterization of the analytes. Section 5.4.2.4 in Chapter 5 discusses MRM and shows how it is superior to SIR in this domain.

The extraction yield (recovery ratio) may be efficiently estimated with an internal standard and more precisely when the chemical structure of the standard is close to that of the analyte. A molecule marked with stable isotopes is strongly recommended, for all the reasons stated previously. The analyte is added to the matrix before sample preparation; the internal standard is added at the end of the preparation at the same concentration as the analyte. After analysis in GC-MS, a comparison of the chromatographic peak areas of the analyte and of the standard gives a correct estimation of the extraction yield.

The two chromatographic peaks are co-eluted and the integrations are made on the ionic currents corresponding to the ions retained for quantification for the analyte and for the standard. The recovery ratio corresponds to the area of the chromatographic peak of the analyte divided by that of the standard and is generally expressed as a percentage. This way of determining a recovery ratio is as accurate as the chromatographic peak area of the analyte is close to that of the standard (recovery yield close to 100%). By proceeding this way, we use a virtual respose function assumed as linear on the given concentration range.

The extraction yield can significantly vary, depending on the concentration of the analyte in the matrix (or on the volume of the sample portion). It is thus recommended to estimate the extraction yield for each analyte on at least two different concentration levels.

Repeatability is estimated from the recovery ratios obtained by reproducing the above procedure four to six times. Repeatability is expressed in the form of a standard deviation. Problems in repeatability are theoretically offset by the use of an internal standard. Experience, however, shows that it is preferable to check that repeatability is satisfactory before starting the method validation to avoid serious problems further along the process.

7.3.4.4   Pre-Validation

After the GC-MS method and sample preparation have been set up and evaluated, one must now consider validating the dosage method to be used. Method validation has progressed a lot in recent years, essentially through the efforts of the pharmaceutical industry. The new validation protocols are much more rigorous than in the past but they require more experiments and thus a lot of time. That is why pre-validation is recommended before validation. This step constitutes a preparatory phase for validation. Essentially, its objective is to determine for each analyte the range of concentrations that makes validation possible. The goal therefore is to reduce the number of analyses by avoiding the injection of samples at concentrations whose validation parameters have no chance of succeeding.

No standard protocol applies to pre-validation. The important point is to determine for each analyte the range of concentrations suitable for validating the method and the concentrations of internal standard to be used.

The goal is to build a response function for each analyte from the temporary limit of quantification (whose value can be increased at the end of the validation) at a concentration fixed by the operator according to the analytical needs or according to the results obtained. It is not necessary to plot the response function for concentrations exceeding the value beyond which the GC-MS response is no longer proportional to the quantity injected. The response functions must be plotted from samples subjected to the complete analysis protocol.

Figure 7.6 presents the procedures for the example presented above concerning the dosage of pesticides in lettuce. Ten lettuces considered as blanks (cultivated in conditions that guarantee a total absence of pesticides) are chopped. Their pieces are mixed to constitute a blank reference matrix from which ten samples of 10 g each are collected. Each of the ten samples is then spiked at a different concentration with a solution that contains all the pesticides. Table 7.1 proposes concentrations for the spiking of pesticides and of standards in each sample.

In Sample 1, each of the pesticides is at a concentration corresponding to its estimated quantification limit: 1.0 ng/g for pesticides 1, 3, 4, and 7; 2.0 ng/g for pesticides 2,

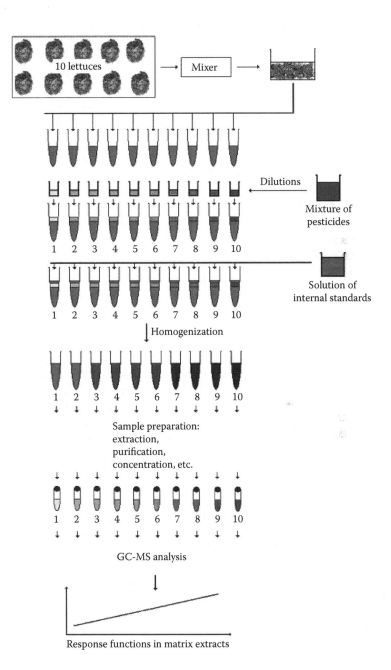

FIGURE 7.6   Setup for quantifying ten pesticides in lettuce.

TABLE 7.1

**Concentrations[a] for Spiking Lettuce with Pesticides
to Develop Quantification Method**

	Sample									
	1	2	3	4	5	6	7	8	9	10
Pesticide 1	1	2	5	10	25	50	75	100	150	200
Pesticide 2	2	4	10	20	50	100	150	200	300	400
Pesticide 3	1	2	5	10	25	50	75	100	150	200
Pesticide 4	1	2	5	10	25	50	75	100	150	200
Pesticide 5	2	4	10	20	50	100	150	200	300	400
Pesticide 6	5	10	25	50	125	250	325	500	750	1000
Pesticide 7	1	2	5	10	25	50	75	100	150	200
Pesticide 8	2	4	10	20	50	100	150	200	300	400
Pesticide 9	5	10	25	50	125	250	325	500	750	1000
Pesticide 10	2	4	10	20	50	100	150	200	300	400
Standard 1	20	20	20	20	20	20	20	20	20	20
Standard 2	30	30	30	30	30	30	30	30	30	30

[a] In nanograms per gram.

5, 8, and 10; and 5.0 ng/g for pesticides 6 and 9. In Sample 10, each pesticide is at a concentration 20 times the LOQ. The internal standards are added to the ten samples at a concentration of 20.0 ng/mL for the first (used for the dosing of the first six chlorinated pesticides) and 30.0 ng/mL for the second (used for the dosing of non-chlorinated pesticides 7 to 10). These values were chosen because they correspond to concentrations for which each standard supplies a chromatographic peak that is easy to integrate precisely (good signal-to-noise ratio) while staying in the low values of the calibration range.

After the range of standard solutions corresponding to the ten samples presented in Table 7.1 is prepared, the samples are injected for GC-MS analysis and the response functions for each pesticide plotted. As an example, Figure 7.7 presents the response function for Pesticide 1. Unlike the response function shown in Figure 7.1, the one in Figure 7.7 was traced from the results obtained from samples subjected to the entire analytical protocol including sample preparation.

At this stage, the development of the analytical method is finished. Pre-validation verifies that response functions are linear or almost linear (Section 7.4.3.1 describes the limitations of quadratic response functions). Repeatability tests, mainly near the provisional LOQ, are performed to ensure that matrix effects have little impact on repeatability. The method is now ready for validation.

7.3.5 ESTIMATING QUANTIFICATION WITH A STANDARD

There are cases where one cannot or does not want to proceed through a complete method development as described next because of urgency issues or because it is not

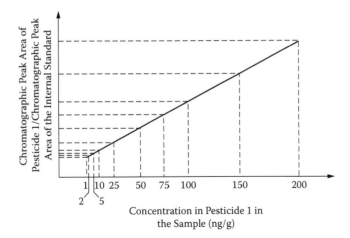

FIGURE 7.7 Response function for Pesticide 1.

necessary for the result of the dosage to be precise. In such cases, an evaluation in a particular analytical context is sufficient. In these conditions, quantification can be done in a way that recalls the one used to estimate the extraction recovery with a deuterated standard. This principle is described in the section dedicated to sample preparation (see Section 7.3.4.1). The difference here relies on the fact that the standard is added to the matrix as soon as the test portion of sample is taken and not after the sample preparation. An estimation of the concentration in analyte can be undergone by comparing the areas of the chromatographic peaks of the analyte and of the standard on the retained ion for the quantification of each of them and by estimating the concentration in analytes according to a rule of three: the concentration in analyte is equal to the concentration in standard multiplied by the area of the peak of the analyte and divided by that of the standard peak. It is obviously imperative for the ions chosen for the analyte and the standard to present the same chemical structure, their m/z ratios differing only by the presence of stable isotopes in the structure of the standard.

By estimating quantification in such a way, the essential aim is to avoid the construction of a response function. The approximate dosage principle relies on the similarities of physico-chemical properties of the analyte and the standard. As mentioned previously, for such a dosage to be acceptable when its approximate character is known, the chromatographic peaks of the internal standard and the analyte must be of approximately the same size and the first must be added in the test sample at a concentration close to that of the second. We are now confronted with a paradox. How can we add the standard to a concentration close to that of the analyte to be dosed if we do not know the concentration of the analyte? If we knew the concentration, we would not seek to dose it! In practice, one or two trials will be required before the areas of the chromatographic peaks will reach relative proportions that are compatible with an empirical dosage.

It is obvious that this procedure is not recommended and should be used only in situations in which no alternative solution is available or as a preparatory step for a more rigorous methodological development. It is in no case applicable to the simultaneous dosing of several molecules because a marked standard in that case cannot be used for dosing of molecules with different physico-chemical properties.

7.4   METHOD VALIDATION

7.4.1   INTRODUCTION

Validation of analytical methods has served as the subject of several books. The descriptions in this section are far from exhaustive! The aim is to explain the concepts of validation to novices in method development because experience with analytical validation methods can influence method set-up. For example, a user confronted with repeatability or robustness issues during an analytical validation can reconsider his choices of sample preparation, ionization mode, or detection mode during the development of a further method.

Method validation is a complex undertaking because there are several ways to validate an analytical method. The validation will generally be based on whether a reference method for the compounds to be dosed already exists. If a reference method exists, the performance of the method under consideration is compared with that of the already validated method. If no reference method exists, a validation procedure coherent with the planned analytical method must be utilized. Proceeding further depends on the context involved, for example, pharmacology, environmental studies, or petrochemistry. As we will see, validation is first of all intended to ensure quality, but it is also subject to some subjectivity.

Some validation protocols use mathematical equations and statistic principles that are relatively complex and that are not discussed here. Computer programs now handle statistical treatment of the data (variance tests, for example). The operator's job essentially consists of entering the experimental results and following a validation protocol established by a specialist in quality procedures.

After discussing the objectives of a validation method and defining important terms, validation will be demonstrated by an example from the field of toxicology: the validation of a GC-MS method to quantify cocaine and its main metabolites in hair.

7.4.2   PURPOSE OF VALIDATION

The purpose of validating an analytical procedure is to demonstrate that it suits the use for which it was conceived. The objective is to dose as exactly and accurately as possible the molecules a laboratory intends to analyze. Validation guarantees the reliability of test results, that is, the measurements carried out with a method are close to "true."

7.4.3   VALIDATION CRITERIA

This section defines the primary method validation criteria. Some concepts, for example, selectivity or response function, were defined in previous chapters and are briefly reiterated below, along with discussions of error and fidelity.

7.4.3.1 Definitions

Specificity—Saying that a method is specific means that it is exclusively dedicated to one or several analytes. In terms of quality, specificity minimizes the risks of dosing a wrong molecule. Note that the concept of zero risk does not exist in analytical chemistry. The characterization of the analyte is based on strict criteria.

Selectivity—A method meets selectivity requirements if dosage results are not disrupted by the presence of molecules other than the analytes—matrix interferents, for example. The selectivity of the method is not expressed as a number. It is evaluated by examining the chromatograms of real sample matrix extracts containing the analytes or spiked with analytes.

Limit of detection (LOD)—The LOD of an analytical procedure is the smallest quantity that can be detected with a given level of confidence but not quantified as an exact value under the experimental conditions described for the procedure.

Limit of quantification (LOQ)—The LOQ of an analytical procedure is the smallest quantity of analyte that can be dosed in a sample under the experimental conditions described for the procedure with defined precision and accuracy.

Response function—This factor translates the relationship between the chromatographic response and the concentration in the substance to be dosed in a given dosing interval. The response function, often confused with linearity (see below), is not obviously linear on the dosing interval (Figure 7.8). Many laboratories validate analytical procedures that include non-linear response functions in particular response functions with quadratic fits. The problem with using a quadratic response function is that the uncertainties of the results of the dosage increase with the measured concentration, as illustrated in Figure 7.9.

Linearity—This is the capacity of an analytical procedure to supply results that are directly proportional to the quantity of substance present in the sample in a given interval of concentrations (Figure 7.10).

The first step of estimating linearity is the dosage of analytes from samples in which their concentrations are precisely known and determining the values of the concentrations by using the response functions traced for each of the analytes. For each compound, the plot of the measured concentrations as a function of the theoretical concentrations must supply a straight line of type $y = ax$ with a slope as close as

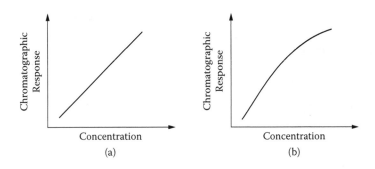

FIGURE 7.8   Linear (a) and non-linear (b) response functions.

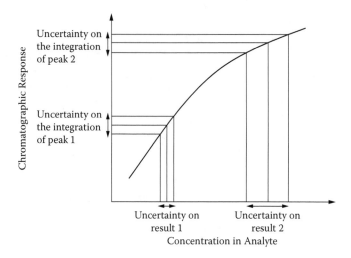

FIGURE 7.9   Uncertainties of dosage results when a quadratic response function is used.

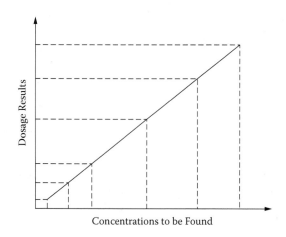

FIGURE 7.10   Linearity of an analytical process.

possible to 1. The linear regression coefficient r^2 associated to this line must also be superior to 0.99, preferably 0.999.

Interval of measure—This is the range of concentrations on which the dosage can be performed with accuracy and precision. It ranges from the LOQ to the highest concentration of the domain of linearity.

7.4.3.2   Error Issues

An analytical result is never perfect. It is always affected by a certain number of errors categorized as systematic (bias) and random errors. For example, an auto-sampler commits systematic errors when it injects 1.1 μL of a sample at a speed of 12 μL.s^{-1} when the specification is to inject a volume of 1.0 μL at 10 μL.s^{-1}.

FIGURE 7.11   Accuracy and precision.

The autosampler errors pervade the whole series of samples. If an operator injects a sample by hand, he commits subjective random errors, for example, by varying the volume injected or the injection speed.

Accuracy—Accuracy expresses the narrowness of the span between the average value obtained from a series of trial results and a value accepted as conventionally true, or as an admitted reference value. Accuracy problems indicate systematic errors.

Precision—This criterion expresses the dispersion of trial results around the average value of the results of these trials. The famous archer figure (Figure 7.11) illustrates the notions of accuracy and precision.

7.4.3.3   Fidelity of Measurement Results

Fidelity in the context of method validation is better known as reproducibility or repeatability. However, each term possesses a precise meaning in terms of quality and fidelity and repeatability should not be confused. The definitions can vary from one field to another and it is important to describe fidelity and repeatability in the documentation of an analytical procedure and detail the conditions of their measure: number of samples analyzed, time lapses between samples, and other factors. The definitions below were used by the European pharmaceutical industry at the time of the writing of this chapter. The criteria linked to the fidelity of measures are generally expressed as coefficients of variation (CVs).

Fidelity—The fidelity of an analytical procedure expresses the narrowness of the accord (measure of dispersion) between the measures obtained from several portions of a single homogeneous sample.

Repeatability—This characteristic is fidelity estimated when the results of independent trials are obtained with the same method on identical portions of a sample in the same laboratory by the same operator, using the same equipment during a short lapse of time. Repeatability is generally estimated over a day and called intra-day repeatability in some of the literature.

Intermediate fidelity—This type of fidelity is estimated when the results of independent trials are obtained with the same method on identical portions of a sample in the same laboratory with different operators, using different equipment over a specific lapse of time. Intermediate fidelity cannot always be estimated. In the case that interests us, that is, GC-MS coupling, an evaluation of intermediate fidelity requires utilizing at least two GC-MS devices and two operators—not possible in all laboratories. In practice, intermediate fidelity is generally estimated like repeatability but over several days and known as between-day repeatability in some of the literature.

Robustness—This is a measure of the ability of an analytical procedure to *not* be affected by minor modifications of factors associated with the procedure. The estimation of the robustness of a method is a subjective process, depending on the context and choice of parameters to be studied determined by the person in charge of the validation. The legitimacy of the choice must be justified during the procedure. One can, for example, study the influence of the pH of the matrix (e.g., urine) during the preparation of the sample on the extraction yield of the analyte. The aim is to verify that a minor modification of pH does not significantly alter the extraction yield. If the room where the samples are prepared is not air conditioned, one can also check that a modification in temperature does not alter the dosage results.

7.4.4 COMPLEMENTARY VALIDATION CRITERIA

The validation criteria enumerated previously are studied in most validation protocols, no matter what analytical method is used. The study of certain complementary criteria, tightly dependant on the analytical context, is often indispensable.

Dilution effect—During a dosage, the estimated quantity of analytes may exceed the maximum concentration of the domain on which the response function was plotted; it is indeed then an estimation since the response function cannot legitimately be used for a precise dosage.

Suppose the range of concentrations used for the plotting of the response function is between 1.0 and 200.0 ng/mL. A first dosage allows estimating that the concentration in analytes in the sample is around 250 ng/mL, extrapolated by prolonging the plotting of the response function beyond the domain of concentrations of the sample. Instead of establishing a new response function and having to re-validate the whole method, the simplest procedure to realize a dosage in this case is to dilute the sample by a factor of two, for example, and repeat the analysis.

By proceeding this way, it is very probable that the result of the integration of the chromatographic peak will be compatible with the use of the validated response function. Proceeding in this manner is legitimate only if one has previously performed tests during method validation to verify that dilution of the sample does not alter the result of the dosage.

Stability in time—This sample parameter must be verified when sample analysis can be delayed for long intervals. This stability criterion is particularly important when dealing with volatile, thermally unstable, or photosensitive compounds. For analysis of biological matrices, the degradation of the matrix (blood, urine, animal or vegetal tissue), although slowed by freezing, is a factor that must be taken into account at the time of method validation. During a toxicological study, for example, part of the sample is generally conserved in a freezer in case the study must be repeated. The repeated test may take place several months after the first one. Under these conditions, the composition of the matrix will change and the efficacy of the sample preparation can therefore be affected.

7.4.5 Application to Real Sample Analysis

After all the validation criteria are established, the application of the method to the analysis of real samples constitutes the ultimate step of the validation process. Up to now, all procedures have been conducted from samples constituted from blank matrices spiked with analytes. The goal now is to verify that the method is applicable to "real" samples. It is indeed possible for the extraction of the analytes from real samples to be less efficient than extraction of samples from a spiked matrix.

Let us consider the examples of human biological samples. It is possible that a molecule to be dosed or its metabolites are subject to physico-chemical interactions with molecules in the matrix. Such interactions may differ if the molecule to be dosed or its metabolites are added to the matrix outside the organism or if they are present in the organism endogenously or after ingestion. In these conditions, the extraction of optimized analytes from spiked matrix samples can be ineffective for real samples.

Clearly, the idea is to verify that the analytes are in fact detected in real samples that we know contain analytes at concentrations above the quantification limits of the method. It is not possible to verify the accuracy of the dosage in this context because we do not know the "real" concentrations that differ from one organism to another.[4]

7.4.6 Example of Method Validation

7.4.6.1 Scientific Context

Examples are generally easier to apprehend than theoretical explanations so this section will present a concrete example of method validation. The described procedure is not universal or exhaustive, but it follows the recommendation of the Société Française des Sciences et Techniques Pharmaceutiques (SFSTP, French Society of Science and Pharmaceutical Techniques) and has proven its efficiency.[5] The number of analyses conducted per day and the number of days during which experiments are conducted are indicative and determined by the person in charge of validation.

The validation method presented here was carried out by the team of C. Staub at the Forensic Laboratory of Geneva University.[6] The point was to validate a method for the dosage of cocaine (COC) and its three main metabolites [ecgonine methylester (EME), anhydroecgonine methylester (AEME), and cocaethylene (COET)] in hair. Their chemical structures are presented in Figure 7.12. The internal standards used were cocaine-d_3 for COC and COET and ecgonin methylester-d_3 for EME and AEME (molecules deuterated on one of the methyl groups).

The method was developed according to a procedure analogous to the one described in Section 7.3.4. Sample preparation consisted of acidic hydrolysis of collected hair, followed by an extraction in solid phase extraction. The analysis of cocaine and its metabolites was carried out in GC-MS with an ion trap in a protocol that associated positive chemical ionization and MS/MS detection for each of the four analytes. No chemical derivation was carried out prior to analysis.

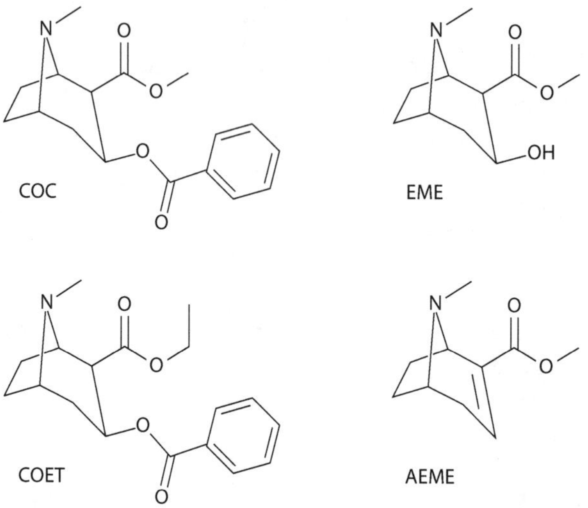

FIGURE 7.12 Chemical structures of cocaine (COC) and its three main metabolites: ecgonine methylester (EME), anhydroecgonine methylester (AEME), and cocaethylene (COET).

7.4.6.2 Validation Process

The validation procedure should include the entire method. The methodology and intermediate results are presented here for cocaine; the process was identical for the three other analytes.

The limit of quantification of the method was estimated at 0.05 ng/mg of hair. The decision was made to attempt to validate the method on a range of concentrations from 0.05 to 5.0 ng/mg. For three days, three analyses were performed daily from samples known as calibrators (samples of matrices spiked with cocaine and metabolites at seven levels of concentration spread over the chosen range). The concentrations of the calibrators and the results obtained for cocaine are given in Table 7.2. The response functions of the dosage method for cocaine were traced (Figure 7.13) from the results presented in Table 7.2. The response function for cocaine was plotted each day with 3 points per concentration (3 injections per day × 3 days).

If a point is *a priori* considered as aberrant because it is very far from the other points of the same abscissa, it is possible to omit it from the plotting of the response function under the condition that one disposes of many points as is the case here and proceeds with a statistical treatment allowing the demonstration of the aberrant character of the point in question.[7] In the presented example, all the measures were taken into account to plot the response functions.

After the response functions are plotted from these analyses, one proceeds with the quality control analysis of the samples. The procedure is again using samples of blank matrices spiked in analytes at concentrations chosen by the user over the whole range of the calibration interval. These samples are analyzed based on the response functions plotted to evaluate the accuracy, repeatability, and intermediate fidelity of the method. The concentrations of the quality controls and the results obtained for cocaine from these samples are given in Table 7.3.

TABLE 7.2
Results Used to Plot the Response Function for Cocaine[a]

Concentration in	Chromatographic Response		
Cocaine (ng/mg)	Day 1	Day 2	Day 3
0.05	0.131	0.145	0.091
0.05	0.119	0.157	0.099
0.05	0.130	0.151	0.092
0.10	0.271	0.298	0.211
0.10	0.282	0.292	0.203
0.10	0.258	0.287	0.213
0.20	0.546	0.599	0.469
0.20	0.427	0.680	0.512
0.20	0.504	0.685	0.457
0.50	1.120	1.289	0.997
0.50	1.087	1.374	0.970
0.50	1.154	1.258	1.024
1.00	2.171	2.457	2.095
1.00	2.202	2.469	2.106
1.00	2.095	2.463	2.300
2.50	5.636	5.019	4.458
2.50	5.916	4.656	4.752
2.50	5.674	5.916	5.227
5.00	12.198	11.133	8.618
5.00	11.781	10.997	9.330
5.00	11.639	13.550	10.140

[a] Results are given without units. Numbers correspond to chromatographic
   peak area of cocaine divided by that of the internal standard.

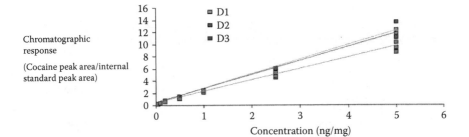

FIGURE 7.13   Response functions plotted for cocaine for each day.

TABLE 7.3

Results of Dosage of Quality Control Samples for Cocaine[a]

Concentration in Cocaine (ng/mg)	Chromatographic Response		
	Day 1	Day 2	Day 3
0.05	0.142	0.153	0.077
0.05	0.180	0.154	0.099
0.05	0.176	0.138	0.094
0.05	0.154	0.140	0.094
0.10	0.257	0.294	0.204
0.10	0.256	0.331	0.174
0.10	0.286	0.292	0.223
0.10	0.270	0.314	0.212
0.50	1.176	1.200	0.983
0.50	1.126	1.179	0.975
0.50	1.135	1.291	1.061
0.50	1.157	1.961	1.041
2.50	5.373	4.600	5.465
2.50	4.732	5.016	4.446
2.50	5.575	5.178	4.755
2.50	5.549	4.900	4.777
4.00	9.262	9.653	7.234
4.00	8.830	6.808	7.413
4.00	10.214	8.066	7.562
4.00	9.461	8.027	7.101

[a] Results are given without units. Numbers correspond to chromatographic peak area of cocaine divided by that of the internal standard.

The linear regression (least squares) method is used to correlate the concentrations determined by the dosage of quality control samples to the theoretical concentrations of these samples.[8] A regression coefficient r2 of 0.999 is obtained by linear regression, thereby demonstrating the linearity of the method on the considered domain of concentrations. Weighting factors can be applied to improve the regression coefficient if necessary.[9]

By applying the values of Table 7.3 to the corresponding response function, it is possible to determine the accuracy (A) of each measure by applying the following relation:

A = (value of determined concentration/value of theoretical concentration) × 100

The dosage of a quality control is considered all the more accurate as the value of A is close to 100%. Table 7.4 gives the determined accuracy for each quality control dosing. Repeatability and intermediate fidelity are estimated for each concentration

TABLE 7.4
Accuracy for Quality Control Dosages

Concentration in Cocaine (ng/mg)	Day 1 Accuracy	Day 2 Accuracy	Day 3 Accuracy
0.05	108.7	94.5	78.5
0.05	141.3	94.6	99.4
0.05	138.2	81.7	95.1
0.05	118.4	83.4	94.6
0.10	108.5	103.9	100.5
0.10	103.7	118.8	85.9
0.10	116.7	103.0	109.7
0.10	110.0	112.1	104.4
0.50	100.8	93.9	95.5
0.50	96.4	92.2	94.6
0.50	97.2	101.3	103.0
0.50	99.2	101.9	101.0
2.50	93.2	103.2	105.8
2.50	82.1	80.4	86.1
2.50	96.7	83.0	92.1
2.50	96.3	78.5	90.2
4.00	100.6	97.0	87.5
4.00	95.9	68.3	89.7
4.00	110.9	81.0	91.5
4.00	102.7	80.6	85.9

Note: For each concentration, the global accuracy is determined as the average of 12 values (4 values per day × 3 days).

and each day by a variance analysis (ANOVA).[10] Tables should be prepared for all concentrations of quality control samples. Table 7.5 gives the results for cocaine at the concentration of 0.5 ng/mg.

7.4.6.3 Results and Conclusions of Validation

Having determined the accuracy, repeatability, and intermediate fidelity for the four analytes dosed by the method for each of the five concentrations, the next step is to construct a table that compiles all the results (see Table 7.6). The acceptance criteria of an analytical method are 20% for accuracy, coefficients of variation less than 15% for repeatability, and less than 25% for intermediate fidelity according to the SFSTP.

By applying these criteria to the results presented in Table 7.6, we can see that the method is validated on the entire domain of concentrations considered for the COC, AEME, and COET analytes. By contrast, the results for EME do not meet the set criteria. Indeed, the accuracy, repeatability, and intermediate fidelity are 40, 58, and 96%, respectively, at a concentration of 0.05 ng/mg. The intermediate fidelity is 26% at 0.10 ng/mg. The repeatability is 16% at 2.5 ng/mg and 17% at 4.0 ng/mg.

TABLE 7.5

Estimation of Repeatability and Intermediate Fidelity for Cocaine at a Concentration of 0.5 ng/mg

	Day 1	Day 2	Day 3
N	4	4	4
Standard deviation (S^2)	2.0%	5.0%	4.2%
Mean value	0.492	0.479	0.493
Repeatability	CV	3.5%	
Intermediate fidelity	CV	3.5%	

TABLE 7.6

Accuracy, Repeatability, and Intermediate Fidelity Results from Quality Control Samples

Compound	Theoretical Concentration (ng/mg)	Measured Concentration (ng/mg)	Accuracy (%)	Repeatability (%)	Intermediate Fidelity (%)
Cocaine (COC)	0.05	0.05	100	11	23
	0.10	0.11	110	7	8
	0.50	0.49	98	3	3
	2.50	2.26	90	6	10
	4.00	3.64	91	9	14
Ecgonine methylester (EME)	0.05	0.02	40	58	96
	0.10	0.12	120	11	26
	0.50	0.44	88	13	14
	2.50	2.00	80	16	16
	4.00	3.45	86	17	23
Anhydroecgonine methylester (AEME)	0.05	0.05	100	11	16
	0.10	0.11	110	15	15
	0.50	0.48	96	11	15
	2.50	2.28	91	7	12
	4.00	3.92	98	3	14
Cocaethylene (COET)	0.05	0.05	100	7	12
	0.10	0.10	100	11	11
	0.50	0.50	100	5	5
	2.50	2.22	89	4	9
	4.00	3.71	93	6	8

Faced with these results, four solutions can be considered. The first involves reconsidering the whole method after having determined the reasons for the validation failure for EME. Was sample preparation not sufficiently selective? Was the MS/MS method badly optimized? This is the most tedious solution that is only of interest if one has managed to determine the nature of the problem. It is indeed useless to reiterate the validation without correcting all the parameters responsible for the failure of the validation procedure. If the choice is to repeat the validation, the process must be performed again for each of the four molecules, even if the three others satisfied the validation criteria. The molecules are in mixture and thus susceptible to interaction. Consequently, they cannot be considered independently if they are the objects of the same analytical method.

The second solution is simply erasing EME from the dosage method and retaining only the three analytes for which the validation was successful. This radical solution is very detrimental to toxicological interpretation. The ecgonine methyl ester is a metabolite whose concentration, compared to that of cocaine, reveals the time of the consumption of the narcotic.

The third solution, satisfactory in the present case, consists of conserving the EME analyte in the method by adapting the validation criteria to the results obtained and reducing the interval of dosage for this compound. One can, for example, fix a detection limit at 0.10 ng/mg and tolerances of 20 and 30%, respectively, for repeatability and intermediate fidelity for EME. This is possible if the values remain coherent with the toxicological context (as is the case here) and under the condition that these criteria are specified in the validation procedure. Validation is first of all an internal quality procedure. Consequently, the tolerances relative to validation criteria are determined by the analysts or the quality manager of the laboratory. Table 7.7 compiles results relative to the validation of the analytical method presented as an example.

TABLE 7.7
Validation Results for Example Analytical Method

Compound	Concentration Range Validated (ng/mg)	Validation Criteria (%)
COC	0.05–5.00	Accuracy 20
		Repeatability 15
		Intermediate fidelity 25
EME	0.10–5.00	Accuracy 20
		Repeatability 20
		Intermediate fidelity 30
AEME	0.05–5.00	Accuracy 20
		Repeatability 15
		Intermediate fidelity 25
COT	0.05–5.00	Accuracy 20
		Repeatability 15
		Intermediate fidelity 25

The fourth solution is no doubt the best: attentively studying the result values of Table 7.6. The accuracy, repeatability, and intermediate fidelity values at the concentration of 0.05 ng/mg are very far from the expected results. This indicates a probable detection problem at that concentration. By erasing the corresponding point in the response function, the slope of the latter will be significantly modified. It is probable that if one uses the new response function for the determination of accuracy, repeatability, and intermediate fidelity values at other concentrations, the new values will be compatible with the validation criteria. The interesting point of this solution compared to the third one is that it is possible to retain the acceptance criteria at 15% for repeatability and 25% for intermediate fidelity for EME.

To provide more credibility, an analytical validation can be subjected to interlaboratory tests. At the end of these tests, each laboratory receives the results (often anonymously) obtained by all the participants and can therefore see whether its own results agree with the average obtained results.

REFERENCES

1. Simpson, N. J. K. 2000. *Solid-Phase Extraction: Principles, Techniques, and Applications*, Boca Raton, FL: CRC Press.
2. Borba da Cunha, A. C., M. J. López de Alda, D. Barceló et al. 2004. Multianalyte determination of different classes of pesticides (acidic, triazines, phenyl ureas, anilines, organophosphates, molinate and propanil) by liquid chromatography-electrospray-tandem mass spectrometry. *Anal. Bioanal. Chem.* 378: 940–954.
3. Lin, D. L., S. M. Wang, C. H. Wu et al. 2008. Chemical derivation for the analysis of drugs by GC-MS, a conceptual review. *J. Food Drug Anal.* 16: 1–10.
4. Llored J. P., Ed. 2012. *Philosophy of Chemistry: Practices, Methodologies and Concepts.* Cambridge: Cambridge Scholars Press.
5. Hubert, P., P. Chiap, J. Crommena et al. 1999. The SFSTP guide on the validation of chromatographic methods for drug bioanalysis. *Anal. Chim. Acta* 391: 135–148.
6. Cognard, E., S. Rudaz, S. Bouchonnet et al. 2005. Analysis of cocaine and three of its metabolites in hair by gas chromatography–mass spectrometry using ion trap detection for CI/MS/MS. *J. Chromatogr. B* 826: 17–25.
7. Schwartz, L. M. 1985. Rejection of a deviant point from a straight-line regression. *Anal. Chim. Acta* 178: 355–359.
8. Chiap, P., A. Ceccato, B. Miralles-Buraglia et al. 2001. Development and validation of an automated method for the liquid chromatographic determination of sotalol in plasma using dialysis and trace enrichment on a cation-exchange pre-column as on-line sample preparation. *J. Pharm. Biomed. Anal.* 24: 801–814.
9. Rudaz, S., S. Souverain, C. Schelling et al. 2003. Development and validation of a heart-cutting liquid chromatography-mass spectrometry method for the determination of process-related substances in cetirizine tablets. *Anal. Chim. Acta.* 492: 271–282.
10. Turner, J. R. and J. F. Thayer. 2001. *Introduction to Analysis of Variance: Design, Analysis, and Interpretation.* Newbury Park, CA: Sage Publications.

8 Spectra Databases for GC-MS

8.1 LIMITATIONS OF MASS SPECTRA INTERPRETATION

Structural elucidation from mass spectra recorded in electron ionization (EI) or chemical ionization (CI) is often difficult, if not impossible, for several reasons. The first is that mass spectra, unlike nuclear magnetic resonance (NMR) spectra, contain too little structural information to allow the determination of the structure of an unknown compound. The second reason is that the fragmentation mechanisms of ions in the gas phase are not always elucidated.

Interpreting spectra requires a minimum of knowledge of organic chemistry. Some knowledge specific to the gas phase and radical ions is required for the interpretation of a mass spectra recorded in electron ionization.[1-3] A few days of training are absolutely necessary if one wants to tackle the interpretation of mass spectra. Practice of a few months or even a few years is indispensable for a good mastering of the related chemistry that may appear atypical.

Consider a comparison with a chess game in which the knowledge of the pieces' moves on the chess board is insufficient to be a good player. Only long experience and a lot of practice with both dissociation mechanisms and the "gymnastics" of spectra interpretation allow a certain degree of confidence. Finally, one must also understand that the interpretation of a mass spectrum of a multifunctional molecule is not instantaneous; it can be a long process requiring several hypotheses. The amount of time required for structural elucidation is often incompatible with the time constraints of an industrial analytical laboratory.

For all these reasons, most users of GC-MS coupling use commercial or other databases to identify the compounds eluted by chromatography. This is obviously possible only if the mass spectra of the molecules to be identified have been previously entered into the system. Chemists who work on the synthesis of new molecules and those interested in degradation products of commercial products, pharmaceuticals, or toxic compound metabolites, for instance, will be confronted with the interpretation of mass spectra as these molecules and their specification data are not available commercially and reference spectra are not producible in a laboratory.

8.2 SPECTRA DATABASES

8.2.1 GENERAL POINTS

A mass spectra database generally contains a library of reference spectra and an algorithm that allows comparisons of target spectra (those to be identified) with spectra contained in the library. Most spectrum data programs for mass spectrometers contain the manufacturers' database research algorithms. If this is the case, the user has the choice between using the equipment algorithm or the one provided by a database library. Although the programs may differ slightly, the choice generally has no impact on the research results.

8.2.2 SPECTRA COMPILATIONS

The two types of mass spectra databases are marketed databases that are sold commercially and specialized self-made databases.

8.2.2.1 Marketed Databases

The marketed spectra databases are sold by institutes, constructors of mass spectrometers, editors on CD-ROM, or by download. Publicity put aside, the two most well known are, by far, the "NIST" database (from the American National Institute of Standard and Technology) and the "WILEY" database (marketed by the famous scientific editor). These are databases that index mass spectra recorded in electron ionization at 70 eV. Generally speaking, the few marketed databases devoted to chemical ionization are not very reliable. With this soft ionization mode, the characteristics of spectra depend greatly on conditions of temperature and pressure within the source. For that reason, spectral reproducibility in CI from one mass spectrometer to another is generally mediocre.

Marketed databases offer the advantage of being operational as soon as they are installed on a computer. They contain tens of thousands of mass spectra—a considerable number even if nature produces several millions of molecules! The indexed spectra obviously correspond to molecules such as pesticides, environmental pollutants, toxins, and drugs that interest the largest number of analysts. Chemists who work in these and other relevant fields are more likely to find spectra of their molecules of interest in marketed databases.

The value of marketed spectra databases is offset by certain inconveniences. Some show more compounds than mass spectra, which means that some compounds were recorded several times. For example, one can find the mass spectra of D-limonene (limonene is a terpene presenting an asymmetrical carbon), L-limonene, and the racemic mixture. The three mass spectra are obviously the same since mass spectrometry (that may not distinguish isomers) does not allow differentiation of enantiomers!

Another disadvantage is that all the spectra compiled are not recorded under standard conditions. The spectra come from many sources and types of analyzers and were recorded on devices whose calibrations are uncertain (is the ionization energy always precisely at 70 eV?). The purities and concentrations of reference molecules are variable. In conclusion, if you use a marketed database, consider the results cautiously.

Some marketed spectra databases are dedicated to specific fields of activity. The database of Robert Adams, for example, indexes about 1200 spectra of molecules used in the perfume and aroma industry that are components of natural essential oils and synthetic compounds.[8] The specificity of this spectra database is that it notes the retention times of the compounds under chromatographic conditions specified by the author along with the mass spectra of relevant molecules. This allows the distinction of terpene isomers by their retention times rather than by their mass spectra. This obviously requires work under GC conditions that are rigorously identical to those described by the editor of the database—same capillary column, carrier gas flow, and programming of oven temperature.

8.2.2.2 Self-Made Spectra Databases

Self-made spectra databases are dedicated exclusively to molecules of interest of the producing laboratory. Recording the spectra of the molecules to be characterized obviously requires having the compounds. The advantage of such a database requires compilation of spectra recorded under "ideal" conditions for future research; the reference spectra should be generated by the same mass spectrometer under the same conditions utilized for target spectra. The differences noted previously for marketed spectra databases do not apply to self-made databases that are compiled to provide better performance than marketed databases.

During the creation of such a spectra database, retention times of the compounds can be included if chromatographic conditions are always the same. This allows identification based on two parameters: spectrum and retention time. The spectra to be entered in the database must first undergo a subtraction of background noise because background noise levels vary from one chromatogram to another according to the nature of matrix interfering compounds, the stationary phase of the capillary column, and its degree of bleeding.

The use of a self-made database does not preclude the use of a marketed spectra database to deal with an unexpected result. The appearance of an unexpected peak in a chromatogram is quite frequent!

8.3 RESEARCH ALGORITHM

The algorithm is the computer program that performs database research. There are several kinds of algorithms but they all utilize the same operational principle. To illustrate the operational principle of the algorithm of the signal treatment program that equips mass spectrometers, the Varian™ algorithm is presented. The database contains 102,386 spectra.

The first step when submitting a target spectrum to database research is pre-selection based on simplified spectra. The target spectrum (from which the chromatographic background noise has been subtracted) is reduced to its 16 most abundant ions. As an example, the simplification of the EI spectrum of α-pinene for the 16 majority ions is shown in Figure 8.1. All the spectra of the database are reduced to 8 of their majority ions.

After the spectra are simplified, the algorithm determines, for each of the 102,386 spectra indexed in the library, how many contain ions in common with the target spectrum. The relative abundances of the ions are not taken into account during this pre-selection step. The results of pre-selection are presented in Figure 8.2. We can see that among the 102,386 spectra reduced to 8 ions, 25 present 8 of the 16 ions of the target spectrum, 167 present 7 of the 16 ions of the target spectrum, and

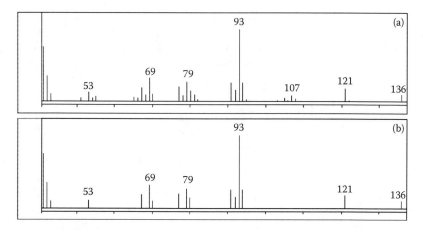

FIGURE 8.1 Reduction of electron ionization mass spectrum of α-pinene (a) to its 16 most abundant ions (b).

Saturn Purity Search Results
Hits Found: 25
Pre-Search Hits Found: 2517
Pre-Search Results

Matches	Count	Total
8	25	25
7	167	192
6	640	832
5	1685	2517
4	3447	5964
3	6483	12447
2	10558	23005
1	20350	43355
0	59031	102386

Saturn Purity Search Parameters
Threshold: 700
Target Ion Range: 45–500
Library MW Range: 0–500
Library Ion Rang: 45–500 (Target Range)

FIGURE 8.2 Results of pre-selection operated by the program on spectra reduced to 8 ions.

640 present 6 of the 16 ions of the target spectrum; 59,031 simplified spectra present no ions in common with the 16 main ions of the target spectrum.

The most similar spectra are selected from a value fixed by the operator (1000 in the chosen example). A total of 2517 simplified spectra present 5 to 8 ions common to those of the target spectrum—1517 more than the 1000 value chosen by the operator. Since we have no reason for ruling out 1517 spectra we therefore select 2517 for the final research. Each of the 2517 spectra will be subjected to a precise comparison with the target spectrum. All the ions of the scanned m/z ratio range are considered and their relative abundances are also taken into account.

O...	*Purity	Fit	RFit	Avg	MW	Name
1	941	955	980	959	122	Phenol, 2,4-dimethyl-
2	940	941	984	955	122	Phenol, 2,3-dimethyl-
3	927	928	972	942	122	Phenol, 2,6-dimethyl-
4	916	922	980	939	122	Phenol, 3,4-dimethyl-
5	911	919	977	936	122	Phenol, 2,5-dimethyl-
6	907	911	958	925	122	Phenol, 3,5-dimethyl-
7	859	879	894	877	122	Benzene, 1-methoxy-4-methyl-
8	858	889	905	884	122	Benzene, 1-methoxy-2-methyl-
9	855	860	909	875	122	Phenol, 3-ethyl-
10	852	857	900	870	122	Benzenemethanol, 4-methyl-
11	846	858	925	877	122	Oxepine, 2,7-dimethyl-
12	843	866	883	864	122	1,3,5-Cycloheptatriene, 1-methoxy-
13	827	864	892	861	122	Bicyclo[3.2.0]hepta-2,6-diene, 5-methoxy-
14	812	908	834	851	122	1-Methoxycycloheptatriene
15	808	817	928	851	164	4-Isopropyl-3,4-dimethylcyclohexa-2,5-dienone
16	802	810	850	821	122	1,3-Cyclopentadiene, 5,5-dimethyl-1-ethyl-

FIGURE 8.3 Results of search on purity in NIST database.

8.4 RESULTS DISPLAY

8.4.1 MATCH CRITERIA

With the algorithm presented in this example, the best solutions supplied by research of the database can be classified according to the criteria of fit, reverse fit, and purity. The fit value translates to the degree of inclusion of a database spectrum in the target spectrum. Reverse fit is based on the opposite principle: the degree of inclusion of the target spectrum in the database. The purity criterion considers the fit and reverse fit values and translates the similarities of the target spectrum and database spectra.

The purity mode is used when a target spectrum is considered "clean" (i.e., not polluted by ions coming from interfering compounds). Figure 8.3 presents the results of purity research in the NIST database. The six best results correspond to six isomers of dimethylphenol. All six have purity indices that range from 907/1000 to 941/1000 (90.7 and 94.1%, respectively). A difference of 3.4% does not allow determination of an isomer considering the spectral repeatability obtained in EI. In this example, it is obviously out of the question to consider that the compound that we are trying to identify is 2,4-dimethylphenol. At best, we can consider that we are dealing with a dimethylphenol but cannot specify which one. Furthermore, we can never totally exclude the possibility that the compound in question is not dimethylphenol and may be another molecule whose mass spectrum is not indexed in the database.

8.4.2 EVIDENCE OF COELUTION

When a purity search does not yield satisfactory results, it may be worthwhile to try another search based on fit. If we obtain a result with a high probability value, it may indicate a chromatographic coelution that was not suspected due to the Gaussian aspect of the peak.

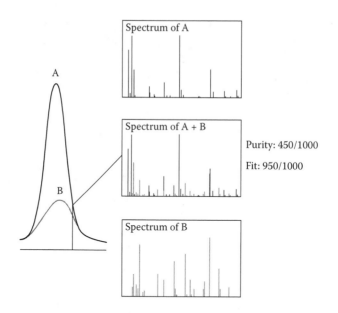

FIGURE 8.4   Chromatographic coelution evidenced by database search in fit mode.

Consider the example presented in Figure 8.4. The compounds A and B are co-eluted. At all points of the chromatographic peak, the mass spectrum results from a superposition of the mass spectra of A and B. Consequently, a search based on purity cannot be efficient because the value of reverse fit is low. If A is predominant, as is the case in Figure 8.4, the value of fit between the target spectrum and that of A is high but the value of the reverse fit is very low due to the presence of the ions of B that prevent the inclusion of the target spectrum in the reference spectrum of A. The correct result for A is therefore accessible through fit and not through purity. After A is identified, it is possible to subtract its mass spectrum from the target spectrum to visualize the spectrum of B. From here, it is possible to carry out a new fit search to attempt to characterize B.

8.5   RESEARCH PARAMETERS

Figure 8.5 shows the window of the program chosen as an example in which the database research parameters are programmed. The window at the top left of Figure 8.5 specifies the classification criteria of the results based on purity, fit, or reverse fit values and the threshold value associated with the criteria above which one wants the corresponding solutions to be shown.

The window below *Library Entry Mol. Weight* shows the range of molecular weights on which the database search is conducted. It is useless to compare the target spectrum to mass spectra of compounds of 1000 amu if one knows that the molecular weight of the compound to be identified cannot exceed 400 amu.

The *Target Ion Range* window allows specification of the range of m/z ratios to be searched. It is important for this range to correspond to that scanned for the

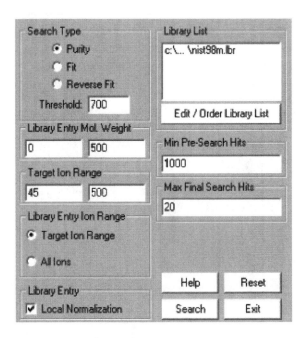

FIGURE 8.5 Window of program in which database research parameters are programmed.

recording of the target spectrum. Otherwise, the search may fail despite the likelihood that the target compound is contained in the database.

The *Library Entry Ion Range* window allows a choice of two options: (1) making the range of m/z ratios to be searched correspond with the values set in the *Target Ion Range* window (the most reliable solution) or (2) considering all the ions present in the spectra of the database independently of the m/z ratios on which the target spectrum was recorded.

The *Library Entry* window at the bottom left of Figure 8.5 allows activation of the *Local Normalization* function. Note that this function is not present in all research algorithms. If it is available, it can be used to perform the research giving preference to the heaviest ions, which are a priori the most characteristic of the target molecule.

The *Library List* window at the top right of Figure 8.5 allows the user to select the mass spectra libraries to be searched. The *Min Pre-Search Hits* window fixes the minimum number of spectra that will be selected at the end of the pre-research phase (see above). Increasing this value expands the length of the research (a small limitation based on the speed of modern computers) but reduces the risks of missing a solution if it exists.

Finally, the *Max Final Search Hits* window fixes the maximum number of results to be shown based on a fit, reverse fit, or purity value above the threshold value fixed by the user. If 80 results are found and the value indicated in the window is 20, only the 20 best results will be shown. If only 3 results based on fit, reverse fit, or purity value exceed the threshold value fixed by the user and the maximum final search hits equal 20, only the 3 best results will be shown.

8.6   DATABASES AND MASS SPECTRA INTERPRETATION

Contrary to popular preconceptions, databases cannot assist in the interpretation of a mass spectrum. As we have just seen, the research algorithms are not based on artificial intelligence. They operate by matching (comparison) of target and reference spectra. Consequently, considering that the molecule suggested by database research is not the right one but should resemble the right one is a delusion that no scientific argument supports. Therefore, two isomers that differ only by the position of a methyl group can supply radically different spectra (often the case with fatty acids, for example).

At the end of a database search, it would be wise to consider only two options: the solution is plausible or it is not. If the result is not plausible, the best next step is to stop using the database and begin the interpretation of the mass spectrum based on chemical criteria (refer to Section 9.2).

8.7   OTHER USES OF DATABASES

In addition to using spectra databases to identify molecules, databases also offer other possibilities.

One can use them to evaluate spectral repeatability, for example, by entering an "average" spectrum compiled from all mass spectra constituting a chromatographic peak, then searching each mass spectrum. One should note for each result the resemblance rate (value for 1000) that indicates the degree of difference between the target and the average spectrum, then determine a standard deviation from all the values. The standard deviation represents a good evaluation of spectral repeatability.

A database also allows instant access to the spectrum of a molecule by searching its name or CAS (Chemical Abstract Service) number. Suppose we must verify immediately that water has not been contaminated by toluene. We plan to use a GC-MS coupling device equipped with a quadrupole. The best way to proceed is by injecting toluene to determine its retention time (under the correct chromatographic conditions) and the main ions of its mass spectrum. One must then analyze the water (after liquid–liquid extraction with an organic solvent or directly by solid phase microextraction coupled with GC-MS) in SIM on the two or three main ions of toluene to reach a limit of detection that is as low as possible. If toluene cannot be injected in the laboratory, quick access to its EI spectrum is available in databases. Ions for SIM acquisition will then be chosen from the reference spectrum.

One can interrogate a database in several ways. For example, a database can be asked to show all the spectra of compounds corresponding to a raw formula or a given molecular weight. One can also make it display mass spectra showing a base peak at m/z 250 and another peak at m/z 300 whose relative abundance ranges from 50 to 60% of that of the base peak.

Certain databases propose programs to help with the interpretation of mass spectra. This is a commercial argument to seduce new users. The programs are totally inefficient for the reasons explained in Chapter 9.

REFERENCES

1. Budzikiewicz, H., C. Djerassi, and D. H. Williams. 1967. *Mass Spectrometry of Organic Compounds*. San Francisco: Holden Day.
2. McLafferty, F. W. and F. Turecek. 1993. *Interpretation of Mass Spectra*, 4th ed. Mill Valley, CA: University Science.
3. Lee, T. A. 1998. *Beginner's Guide to Mass Spectral Interpretation*. New York: John Wiley & Sons.
4. NIST Standard Reference Data. 2012. http://www.nist.gov/srd/nist1a.cfm.
5. McLafferty, F. W. 1989. *Wiley Registry of Mass Spectral Data*, 5th ed. New York: John Wiley & Sons.
6. Scientific Instrument Services. Wiley Registry™ of Mass Spectral Data, 9th ed. (and other specialty spectral libraries). http://www.sisweb.com/software/ms/wiley.htm
7. Milman, B. S. 2005. Identification of chemical compounds. *Trends Anal. Chem.* 24: 493–508.
8. Adams, R. P. 2001. *Identification of Essential Oils Components by Gas Chromatography/ Quadrupole Mass Spectrometry*. Carol Stream, IL: Allured Publishing.

9 Introduction to Mass Spectra Interpretation

9.1 INTRODUCTION

The interpretation of a mass spectrum cannot be improvised: experience shows that chemists who are not properly trained in mass spectrometry are often embarrassed when they start tackling the interpretation of mass spectra. This is because the chemistry of ions in the gaseous phase is often different from the chemistry of the same ions in solution. Most chemists learned to deal with solutions using techniques such as pH metrics, titration, and synthesis. Furthermore, except for photochemistry, classical training treats species with even numbers of electrons: molecules and MH$^+$ or [M-H]$^-$ ions, respectively, emitted from protonated or deprotonated molecules.

Electron ionization—by far the most used ionization mode in GC-MS—produces radical ions with odd numbers of electrons whose reactivities are very different from those of ions with even numbers of electrons. Consequently, some training is required to interpret mass spectra correctly.

A book by this author (in French) dedicated to the interpretation of mass spectra in GC-MS supplies many exercises and solutions. Practice is indispensable for mastering spectra interpretation.[1]

This chapter is composed of three parts. The first covers principles of thermochemistry that are fundamental to the comprehension of the mechanisms involved in spectral interpretation: electronegativity, chemical bonds, acidity and basicity, inductive and mesomeric effects, the Audier-Stevenson rule, and stability rules for radicals in the gas phase.

The second part of this chapter discusses the formation and fragmentation of ions according to the ionization modes used in GC-MS coupling: electron ionization and positive and negative chemical ionization. The dissociation mechanisms of ions (simple cleavages, rearrangements, and secondary fragmentation) are described in detail.

The final section suggests a strategy for the interpretation of mass spectra.

9.2 CHEMISTRY PRINCIPLES FUNDAMENTAL TO MASS SPECTRA INTERPRETATION

9.2.1 ELECTRONEGATIVITY

9.2.1.1 Definition

Electronegativity is an intrinsic property of each atomic element. It determines the affinity of the element for electrons and therefore its capacity to retain, release, or

1	2	3	4	5	6	7	8	9	10	11	12	13	14	15	16	17	18
H 2.2																	He
Li 1.0	Be 1.6											B 2.0	C 2.5	N 3.0	O 3.4	F 4.0	Ne
Na 0.9	Mg 1.3											Al 1.6	Si 1.9	P 2.2	S 2.6	Cl 3.2	Ar
K 0.8	Ca 1.0	Sc 1.4	Ti 1.5	V 1.6	Cr 1.7	Mn 1.5	Fe 1.8	Co 1.9	Ni 1.9	Cu 1.9	Zn 1.6	Ga 1.8	Ge 2.0	As 2.2	Se 2.5	Br 3.0	Kr
Rb 0.8	Sr 0.9	Y 1.2	Zr 1.3	Nb 1.6	Mo 2.2	Tc 2.1	Ru 2.2	Rh 2.3	Pd 2.2	Ag 1.9	Cd 1.7	In 1.8	Sn 2.0	Sb 2.0	Te 2.1	I 2.7	Xe 2.6
Cs 0.8	Ba 0.9		Hf 1.3	Ta 1.5	W 1.7	Re 1.9	Os 2.2	Ir 2.2	Pt 2.2	Au 2.4	Hg 1.9	Ti 1.8	Pb 1.8	Bi 1.9	Po 2.0	At 2.2	Rn
Fr 0.7	Ra 09																

La 1.1	Ce 1.1	Pr 1.1	Nd 1.1	Pm 1.1	Sm 1.2	Eu 1.2	Gd 1.2	Tb 1.2	Dy 1.2	Ho 1.2	Er 1.2	Tm 1.2	Yb 1.1	Lu 1.0
Ac 1.1	Th 1.3	Pa 1.5	U 1.7	Np 1.3	Pu 1.3	Am 1.3	Cm 1.3	Bk 1.3	Es 1.3	Fm 1.3	Md 1.3	Md 1.3	No 1.3	Lr 1.3

FIGURE 9.1 Values of electronegativities of atoms according to Pauling scale.

scavenge them. Electronegativity depends on the number of electrons that constitute the electronic layer of the atom that is furthest from the core (valence layer). One can appreciate the electronegativity of an element by referring to its position in the periodic table established by Mendeleev.

Among the different scales of electronegativity, that of Pauling is the most frequently used. Figure 9.1 gives the electronegativity values of atoms according to the Pauling scale.[2] Electronegativity increases from left to right along a period (line) of the table since the number of protons increases with the atomic number. As the positive charge of the core increases, the attraction exerted by the core on the electrons also increases.

Electronegativity decreases from top to bottom along a family (column of the table) since the addition of an electronic layer distances the valence electrons from the core and the screen effect exerted by the internal layer electrons is accentuated. In terms of spectra interpretation, the first approximation concerns the most common atoms (H, C, O, N, Si, P, S, F, Cl, Br, I) among the molecules classically analyzed by GC-MS. Atoms from the same column of the Mendeleev table will present relatively similar chemical properties.

9.2.1.2   Electronegativity and Chemical Bonds

A chemical bond is a pooling of electrons between two atoms. The trigger of this pooling is the stabilization of atoms by saturation in electrons of their valence layers. The difference of electronegativity between two atoms determines the nature and force of their connection. Among the several kinds of chemical bonds, covalent bonds and ionic bonds are usually used as references.

In a covalent bond, the electronic pair that constitutes the bond is evenly spread between the two elements only if the atoms are identical and possess the same electronegativity. In this case, the two atoms are electrically neutral and the bond is not polarized. If the two atoms involved in a bond are different, the most electronegative atom attracts the electrons and is partially negatively charged; the least electronegative atom is partially positively charged. This is a polarized bond. The greater the difference in electronegativity between the two atoms, the more the bond is polarized.

The ionic bond is the most polarized; the electronic pair is almost entirely located on the most electronegative of the two elements, which is why we speak of ions rather than atoms. The bond arises from Coulomb attraction between ions.

As an example, consider the H_2, H_2O, and NaCl molecules. H_2 hydrogen molecule is apolar. The H-H bond is not polarized because it consists of two identical atoms of the same electronegativity. The electrons of the H-H bond are spread symmetrically between the two atoms. The H_2O water molecule is polar because the O-H bonds are strongly polarized due to the difference in electronegativity between the oxygen (3.4) and hydrogen (2.2) atoms. The distribution of electrons constituting the O-H bond is asymmetrical; they are closer to oxygen than to hydrogen. Sodium chloride is in fact an ionic complex between a cation (Na^+) and an anion (Cl^-). The electronegativity values of chlorine and sodium, respectively, are 3.2 and 0.9 on the Pauling scale.

9.2.2   INDUCTIVE EFFECT

9.2.2.1   Definition

Polarization of chemical bonds described above is at the origin of the inductive effect. Inductive effect is either an electron-attractive power (written as +I) or an electron-repulsive power (–I) of an atom or group of atoms within a molecule or an ion.

An atom or group of atoms of weak electronegativity exerts an electron-releasing inductive effect whereas an atom or atom group of high electronegativity exerts an electron-withdrawing inductive effect. To clarify, it is generally considered that the hydrogen atom exerts no effect, the atoms of the carbon column (C, Si) have an electron-releasing effect, and the atoms N, P, O, S, and the halogens (F, Cl, Br, I) have an electron-withdrawing effect.

Consider the case of ions because they are of interest in mass spectrometry. Inductive effects take part in the stabilization or destabilization of ions. Cations are stabilized by groups exerting an electron-releasing inductive effect (mainly alkyl groups) and destabilized by chemical groups exerting an electron-withdrawing inductive effect: halogens and halogenated alkyl groups, amines, alcohols, ethers,

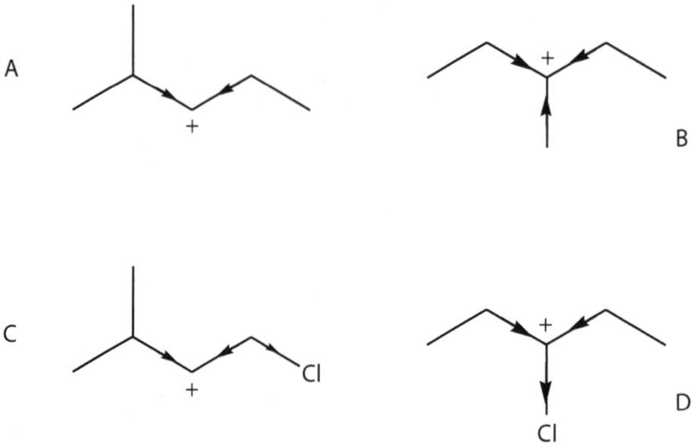

FIGURE 9.2   Direct inductive effects in ions A, B, C, and D.

unsaturated carbon chains (aromatic cycles in particular), and others. Obviously, the opposite effect occurs with anions.

The inductive effect is classically represented by an arrow on the concerned bond or bonds within an ion; the arrow is oriented toward the electron-withdrawing group. The inductive effect spreads along the chemical bonds but its influence on ion stability quickly decreases with the distance between the bond and the charge. Four ions and their most intense inductive effects are represented in Figure 9.2

Cation B is stabilized better than cation A because B is stabilized by three direct electron-releasing inductive effects versus two for A. Cation C is stabilized better than cation D. While both benefit from two electron-releasing inductive effects from alkyl groups adjacent to the charge, D is destabilized by the electron-withdrawing inductive effect exerted by the chlorine atom adjacent to the charge. Note that in cation C the chlorine atom also contributes to its destabilization since it exerts an electron-withdrawing inductive effect on the carbon atom to which it is bonded. Consequently, the carbon is depleted of electrons and the electron-releasing inductive effect that it can exert on the carbon carrying the charge is considerably diminished. Nevertheless, the chlorine atom is further from the charge in C than in D and its destabilizing effect for the cation is weaker than for D.

9.2.2.2   Inductive Effects and Mass Spectrometry

In mass spectrometry, the knowledge of inductive effects allows an analyst to understand (and perhaps anticipate) the evolution of an ion. As an example, Figure 9.3 shows the transfer of a hydride ion (H^-) on a carbon initially carrying a positive charge. Such a transfer is thermodynamically favorable since it transforms a monosubstituted carbocation weakly stabilized by an alkyl chain into a very stable carbocation trisubstituted by alkyl groups with electron-releasing inductive effects.

Transfers of hydride ions are frequent in mass spectrometry because they require only very weak activation energy (see Section 9.2.7). With the exception of CH_3^+, no primary carbocation exists in the gas phase. The $C_2H_5^+$ cation presents a symmetrical geometry corresponding to an ethylene bridged by a proton, that is, the charge is not directly carried by one of the two carbon atoms; it is spread equally

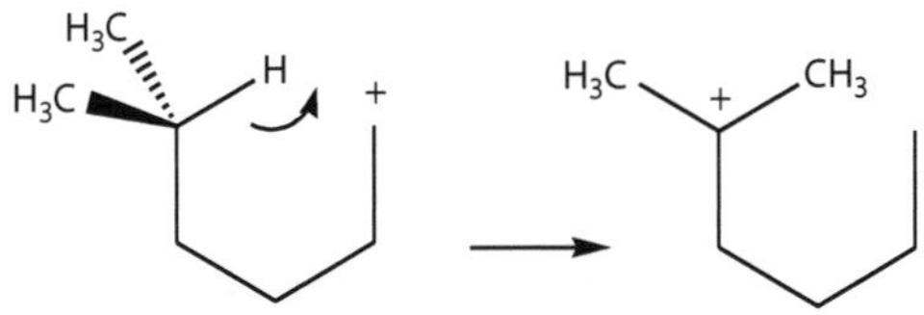

FIGURE 9.3 Example of hydride transfer leading to a more substituted (and therefore more stable) carbocation.

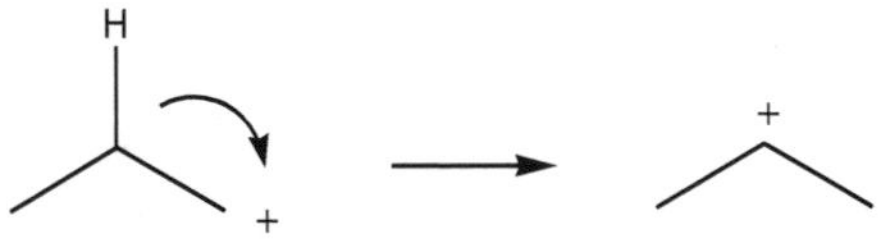

FIGURE 9.4 Spontaneous rearrangement of primary propyl cation into secondary propyl cation in gas phase.

between them. Figure 9.4 illustrates the spontaneous rearrangement of a primary propyl cation into a secondary propyl cation that is stabilized better by electron-releasing inductive effects.

9.2.3 Mesomeric Effects

9.2.3.1 Definition

Mesomery is the delocalization of certain electrons on several atoms of a molecule. This delocalization is generally referred to as resonance or conjugation. It is accompanied by an energetic stabilization of the chemical structure. The mesomeric effect concerns pi (π) electrons, pairs of free electrons, and charges. The cases of benzene and butadiene are often used as examples to explain the phenomenon of conjugation.

9.2.3.2 Examples: Benzene and Butadiene

Benzene (C_6H_6) is a planar molecule that possesses six identical carbon-carbon bonds. One can write this molecule in two equivalent ways called limit forms (Figure 9.5). The passage from one limit form to another is indicated by a double headed arrow ($\leftrightarrow$).

In reality, the π electrons are delocalized in a homogeneous manner on the six carbon atoms. Therefore, neither of the two limit representations in Figure 9.5 corresponds to reality. In the left side representation, the C_1-C_2 bond is double; it is a

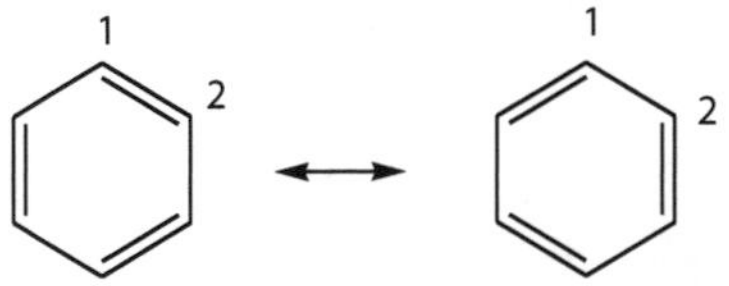

FIGURE 9.5 Limit forms of benzene.

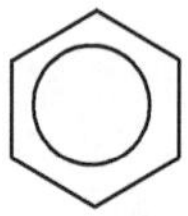

FIGURE 9.6   Resonance hybrid of the molecule of benzene.

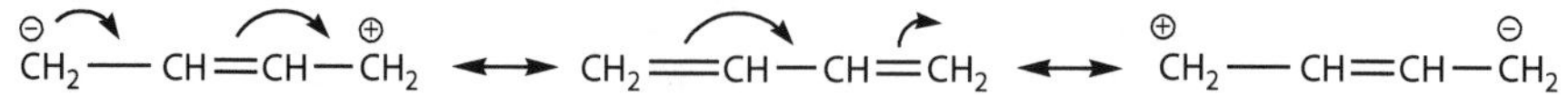

$$\overset{\ominus}{CH_2}\text{—}CH\!=\!CH\text{—}\overset{\oplus}{CH_2} \;\longleftrightarrow\; CH_2\!=\!CH\text{—}CH\!=\!CH_2 \;\longleftrightarrow\; \overset{\oplus}{CH_2}\text{—}CH\!=\!CH\text{—}\overset{\ominus}{CH_2}$$

FIGURE 9.7   Limit forms of 1,3-butadiene.

simple bond in the right side representation. In reality, this bond is neither simple (two electrons) nor double (four electrons); it is a hybrid bond. This is also the case for the other carbon–carbon bonds of the benzene molecule. The representation that best reflects reality is known as the resonance hybrid (Figure 9.6). The circle in the center ring indicates that the electrons are perfectly delocalized between the six carbon atoms. Although pertinent, this representation is rarely used in the writing of reaction mechanisms involving the aromatic ring because it does not allow clear depiction of the electron transfers involved.

The molecule of 1,3-butadiene is also often used as an example to explain the conjugation phenomenon. This molecule can be described by three limit forms, two of which make the charges appear. These limit forms are presented in Figure 9.7.

9.2.3.3   Mesomeric Effects and Cations

In mass spectrometry, whether the mesomeric effect will or will not stabilize the formed ions is of interest. In the examples above, we see that conjugation consists of a delocalization of electrons. For ions, this electron delocalization is revealed by a delocalization of the charge. Consider cations A and B in Figure 9.8. Cation A is not stabilized by resonance and cation B is conjugated. The curved arrow depicts the move of the π electrons of the double bond onto the adjacent bond to constitute another double bond. As the resonance hybrid of B presented at right demonstrates, the positive charge is distributed homogeneously between carbon atoms 1 and 3. This charge distribution is accompanied by a stabilization of the cation. A consequence of this phenomenon is the frequent observation of eliminations from alkyl carbocations leading to a carbocation conjugated with a double bond and therefore more stable (see Section 9.3.3.2).

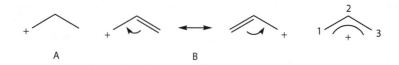

FIGURE 9.8   Cation A (charge non-conjugated), limit forms of cation B (charge conjugated), and resonance hybrid of B.

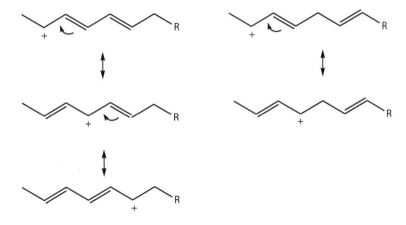

FIGURE 9.9 Resonant stabilization of isomeric carbocations.

We now consider the two isomeric carbocations of Figure 9.9. At left, the charge is delocalized on three carbon atoms. In the right structure, the delocalization is operated on only two carbon atoms (the double bonds are not conjugated). We can conclude that the left carbocation is more stable than the right one.

We have seen in the examples above that π electrons can allow charge delocalization and thus the stabilization of carbocations. This is also the case for n electrons of free electron pairs of heteroatoms. Figure 9.10 illustrates the stabilization of cations by mesomeric effects involving n electrons of heteroatoms.

9.2.3.4 Mesomeric Effects and Anions

The previous examples of mesomeric effects involved carbocations. Carbanions are not studied as intensively as carbocations in GC-MS coupling. Nevertheless, the stabilization of anions by mesomeric effect is also possible and relies on the same electron delocalization principle. Figure 9.11 illustrates the stabilization of anions by mesomeric effect.

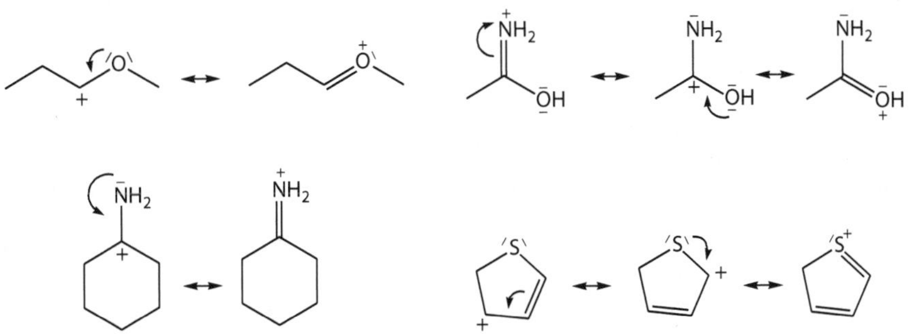

FIGURE 9.10 Examples of stabilization of cations by mesomeric effects involving n electrons of heteroatoms.

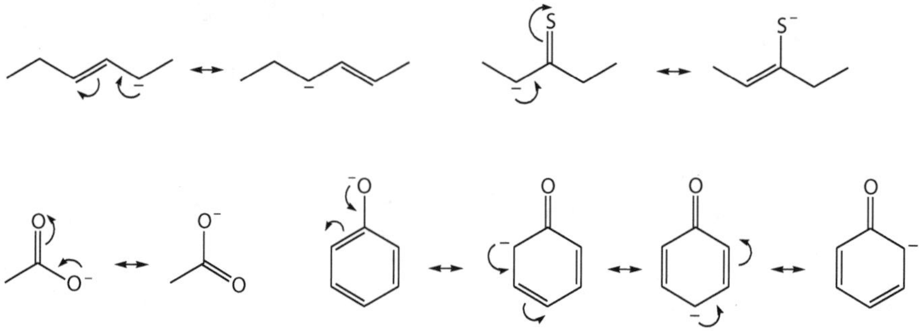

FIGURE 9.11 Examples of anions stabilized by mesomeric effect.

9.2.3.5 Mesomeric Effects and Radicals

The most commonly used ionization mode in GC-MS is electron ionization (EI). EI forms radical molecular ions $M^{+\bullet}$ (ions with odd numbers of electrons) that generally fragment themselves abundantly to render ions with even numbers of electrons and ions with odd numbers of electrons. The interpretation of mass spectra recorded in EI therefore implies the study of radical species.

The mesomeric effect applies to radicals, neutrals, cations, and anions. With radicals, the electron transfers between limit forms are written with half arrows to signify that a single ion participates in the conjugation. Figure 9.12 presents the two limit forms of a conjugated butenyl radical and the corresponding resonance hybrid. The radical is delocalized between carbon atoms 2 and 4.

In the case of a radical cation, the mesomeric effects can involve the charge and/or the radical as is the case in Figure 9.13. A vinyl ether ionized by electron ionization is

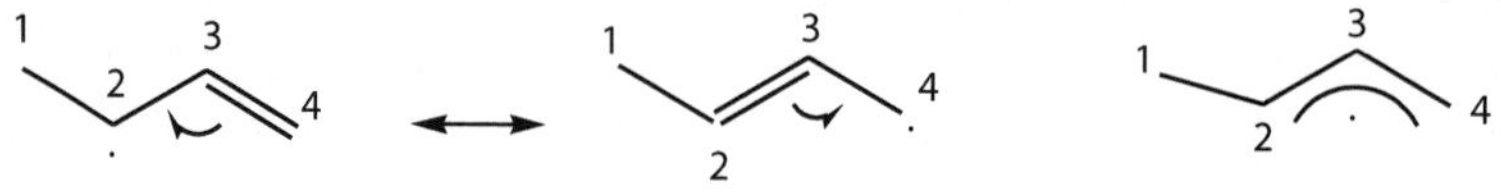

FIGURE 9.12 Two limit forms of a conjugated butenyl radical and corresponding resonance hybrid.

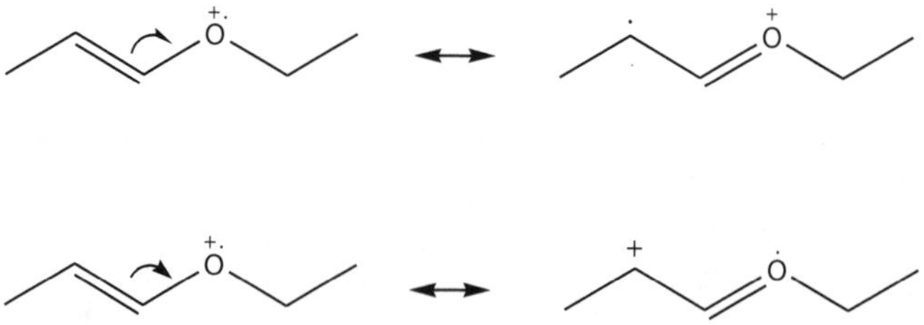

FIGURE 9.13 Limit forms of an ionized vinyl ether stabilized by resonance. Top: limit forms involving delocalization of the radical (single arrow mechanism). Bottom: limit forms involving delocalization of the charge (double arrow mechanism).

stabilized by the delocalization of the radical (top) and by that of the charge (bottom) on two atoms (one carbon and the oxygen atom).

9.2.4 IONIZATION POTENTIAL AND IONIZATION SITES

The ionization potential (IP) is the energy that must be supplied to remove an electron from a molecule (see Chapter 3). IP value tables are available in the literature.[3] Table 9.1 gives the values of ionization potentials of several organic compounds.[4,5]

In GC-MS coupling, the electron ionization technique removes an electron from the eluted molecules of the chromatograph. Generally speaking, the electrons that are the easiest to remove are n electrons, free electron pairs of heteroatoms (N, O, S, P, halogens, etc), and π electrons of double and triple bonds. The sigma (σ) electrons constituting simple bonds are the most difficult to remove. That is why hydrocarbons composed of exclusively σ electrons are not ionized as efficiently as other molecules.

A molecule often presents several ionization sites. When a site presents an ionization potential far inferior to those of the other sites, there is no ambiguity. The electron is removed exclusively from this site and all the ions of the mass spectrum can be interpreted as resulting from dissociations from a unique type of molecular ion $M^{+\cdot}$. For instance, in the case of n-butanamine (Figure 9.14), it is far easier from an energetic view to remove an electron n from the nitrogen atom than an electron σ from one of the C-C, C-H, C-N, or N-H bonds. Most of the ions observed in the mass spectrum can be interpreted by postulating that ionization occurred exclusively on the nitrogen atom.

By contrast, when a molecule presents several ionization sites that are in competition, as is the case with alachlor (Figure 9.15), EI will produce within the source of the mass spectrometer a mixture of $M^{+\cdot}$ ions carrying the charge on several sites. One would generally have to consider these different sites to achieve the interpretation of all the ions of the mass spectrum.[6] To indicate that a molecule is ionized without specifying the localization of the charge, we adopt the method in Figure 9.16. The half bracket followed by a plus sign (+) and a period (.) on the top right means that an electron was removed from the system and that we are in the presence of a radical ion.

TABLE 9.1
Ionization Potential Values of Several Organic Molecules

Molecule	IP (eV)	Molecule	IP (eV)	Molecule	IP (eV)
Pentane	10.4	Benzene	9.2	Ethyl propanoate	10.0
Decane	9.7	Phenol	8.5	Acetic acid	10.7
Hexylamine	8.6	Aniline	7.7	Thiophenol	8.3
Hexanol	9.9	Pyridine	9.3	Acetic anhydride	10.0

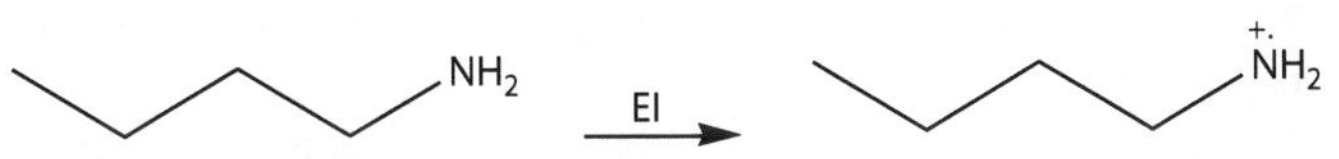

FIGURE 9.14 Electron ionization of n-butanamine.

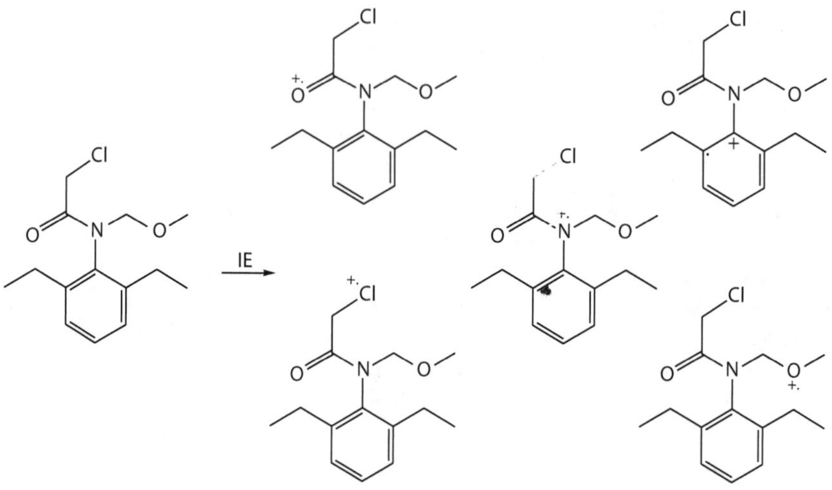

FIGURE 9.15 Ionization sites of alachlor.

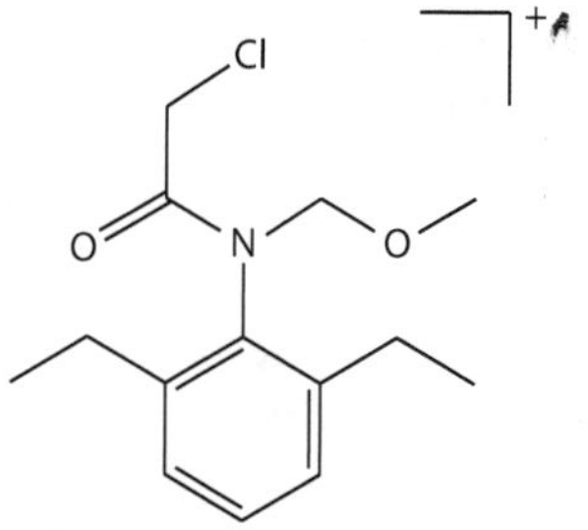

FIGURE 9.16 Indication that a molecule is ionized without specifying localization of the charge.

9.2.5   ACIDITY, BASICITY, PROTON AND ELECTRON AFFINITIES

9.2.5.1   Acidity and Basicity

The concepts of acidity and basicity are usually defined for aqueous solutions but they are broadly applicable to chemistry in the gas phase. According to the definition by Brönsted and Lowry, the acidity of a molecule is its capacity to give a proton; its basicity is its capacity to capture a proton.[7]

In mass spectrometry, acidity and basicity are both of interest in relation to chemical ionization. First, the comparison of acidities and basicities relative to the analyte and a reagent gas will allow an estimation of the feasibility of ionization (refer to Chapter 3). Second, understanding the acidities and basicities relative to the different ionization sites of an analyte is a great help for the interpretation of the resulting mass spectrum.

Therefore, in the case of a multifunctional molecule such as penthidinic acid, one can predict that protonation in positive chemical ionization will occur mainly on the

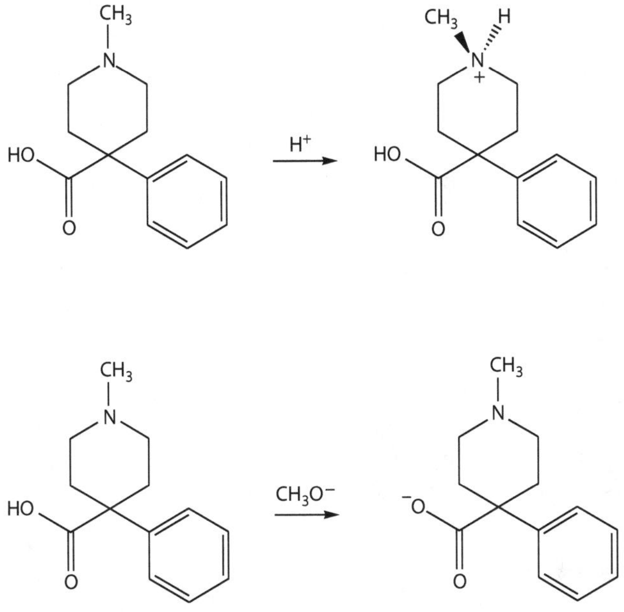

FIGURE 9.17 Main protonation and deprotonation sites of penthidinic acid.

tertiary amine that is the most basic chemical group of the molecule. In negative chemical ionization, deprotonation in the presence of a base such as the methanolate anion will take place mainly at the level of the carboxylic acid, which is the most acid function of the molecule (Figure 9.17).

9.2.5.2 Proton Affinity

In mass spectrometry, the proton affinity parameter is used more than basicity. The proton affinity of a molecule M is the standard enthalpy of reaction [designated delta H (ΔH)] associated with the reaction $MH^+ \rightarrow M + H^+$ in the gas phase.[8] The higher the ΔH value, the more energy that must be supplied to remove a proton from M. The basicity of M corresponds to ΔG (standard free enthalpy of reaction) of the same reaction. Thus, $\Delta G = \Delta H - T\Delta S$, where T is the temperature of the system and ΔS is the entropy variation. ΔS is generally difficult to determine. For that reason, mass spectrometrists generally utilize proton affinities. Table 3.1 in Chapter 3 lists the proton affinity values of the main families of organic compounds.

9.2.5.3 Electron Affinity

In mass spectrometry, electron affinity (EA) is a useful factor for working in electron attachment, as we will see in Section 9.5.1. By definition, electron affinity is the energy that must be supplied to a negative ion to remove an electron from it. It corresponds to the ΔH of the reaction $X^- \rightarrow X + \bar{e}$. Note that in this case $\Delta H = \Delta G$ because the variation of entropy ΔS of the system is nil. To estimate the feasibility of the capture of an electron by a molecule, one must consider the electron affinity of the corresponding negative ion; the higher the affinity is, the easier the capture of an

TABLE 9.2

Electron Affinities of Selected Organic Anions

Anion	EA (kJ/mol)	Anion	EA (kJ/mol)	Anion	EA (kJ/mol)
F_2^-	10.4	Cl_2^-	9.2	$C_6F_6^-$	10.0
CH_3O^-	9.7	$C_6H_6^-$	8.5	$C_{10}H_8^-$ (azulene)	10.0
CH_3S^-	8.6	$CH_3COCH_2^-$	7.7	$C_6H_5NO_2^-$	10.7

Source: Harrison, A. G. 1992. *Chemical Ionization Mass Spectrometry.* Boca Raton, FL: CRC Press. With permission.

electron. Table 9.2 lists the electron affinity values of several organic anions.[8] The electron affinity values of halides are very high due to the high electronegativities of halogens. That is why negative chemical ionization represents a technique of choice for the analysis of these molecules.[9]

9.2.6   AUDIER-STEVENSON RULE

The Audier-Stevenson rule allows an analyst to predict which "side" of the bond will keep the charge in case of bond cleavage. Consider an ionized molecule A-B⁺·. One must compare the ionization potentials of the radicals A· and B· to establish which of A or B will carry the charge in the case of the cleavage of the A-B bond. The charge will be carried by the entity whose ionization potential is the lowest.

Figure 9.18 illustrates the Audier-Stevenson rule in the case of ionized acetone. Will the rupture of the $CH_3OC\text{-}CH_3$ bond lead to the CH_3CO^+ ion and to the radical $CH_3^\cdot$ (pathway a) or to the CH_3^+ ion and to the radical $CH_3CO\cdot$ (pathway b)? The ionization potential of the radical $CH_3CO\cdot$ (680 kJ/mol) is inferior to that of the radical $CH_3^\cdot$ (946 kJ/mol) so the latter will be ionized.

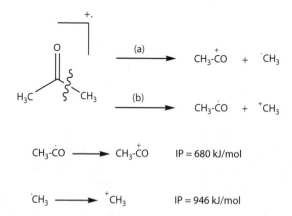

FIGURE 9.18   Audier-Stevenson rule applied to ionized acetone.

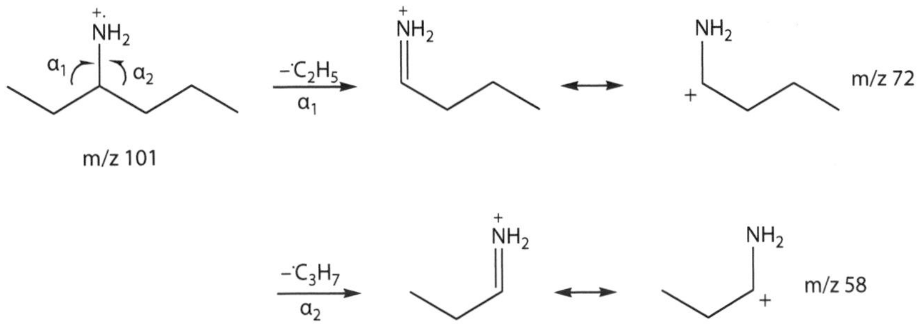

FIGURE 9.19 The α cleavages from ionized 3-hexanamine.

9.2.7 STABILITY OF RADICAL SPECIES

Only ions appear in a mass spectrum. However, the fragmentations of molecular ions also produce neutral species, molecules, and radicals that are not detected but participate nevertheless in the thermochemistry of dissociation reactions. Figure 9.19 shows the alpha cleavages (see Section 9.3.1) possible from 3-hexanamine ionized by electron ionization.

One cleavage leads to the formation of an m/z 72 ion (α_1), the other to the formation of an m/z 58 ion (α_2). The m/z 72 ion is stabilized better than the m/z 58 ion because the electron-releasing inductive effect of the isopropyl group is superior to that of the ethyl group. One must not deduce that the formation of the m/z 72 ion is favored. This amounts to neglecting the stability of the radicals formed during these alpha ruptures.

The stability of radicals follows that of the carbocations since the most substituted or largest radical is the most stable. In the example of 3-hexanamine, the α_1 rupture leads to a more stable ion (m/z 72) and to a less stable ethyl radical than α_2 (m/z 58 ion and propyl radical). In the competition between two fragmentations, the stability of radicals wins over that of cations. Consequently, in the example of ionized 3-hexanamine, one can predict that the α_2 cleavage will be favored (Figure 9.19).

9.3 DISSOCIATION MECHANISMS AND INTERPRETATION OF MASS SPECTRA IN ELECTRON IONIZATION

9.3.1 SIMPLE CLEAVAGES

The two types of simple cleavages from the molecular ions in EI are α (alpha) cleavage and σ (sigma) cleavage. The electronegativity of the ionized atom determines the nature of the simple cleavage usually observed.

9.3.1.1 Alpha Cleavage

Cleavage α is certainly by far the most common mechanism related to the fragmentation of M$^{+\bullet}$ ions. It is a homolytic bond cleavage; that is, the two electrons that constitute the bond separate and each rests on a different atom.

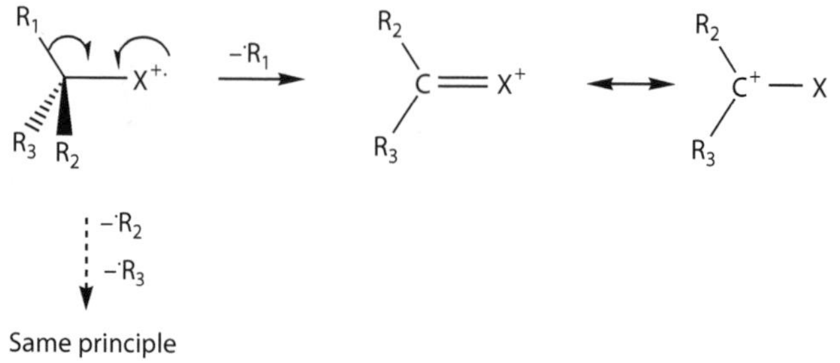

Same principle

FIGURE 9.20 General mechanism of α cleavage.

When an atom that carries a free pair of electrons (O, N, S, P, e.g.,) is ionized, one of the two electrons of the pair is ripped off. The reactivity of the resulting ion is induced primarily by the radical (and not by the charge as beginner spectrometrists often think). Because the radicals are very unstable, the electron that is single after the ionization tries to pair with a neighbor electron. The closest electron susceptible to pairing is one of those that compose the alpha bonds because the bond adjacent with the charged atom is, in fact, electronically impoverished and weakly inclined to give off an electron.

Figure 9.20 illustrates the general principle of α cleavage. Note the simple arrows that indicate the involvement of a single electron. The pairing of two electrons leads to the formation of a double bond; the charge is still carried by the initially ionized X atom. In a mesomeric structure, the positive charge is carried by the carbon atom linked to X. An α cleavage systematically leads to the formation of an ion with an even number of electrons and the elimination of a radical. The possible evolutions of the formed ion are presented later in this chapter. With the exceptions of specific substituents, there are as many possible α cleavages as there are α substituents (three substituents are designated R_1, R_2, and R_3 on Figure 9.20).

Figure 9.21 displays the electron ionization spectrum of n-ethyl,1-propanamine and the different α cleavages possible from the molecular ion of the latter. Each α cleavage is presented as a single arrow from an α bond in the direction of the radical to describe the electron reorganization and the resulting fragmentation. The notation with two simple arrows adopted in Figure 9.21 is recommended for beginners as a reminder that the radical (and not the charge) induces the fragmentation.

Note in the case of n-ethyl,1-propanamine that four α cleavages are possible including the losses of methyl and ethyl radicals that correspond to the break of alkyl groups on one hand and the elimination of a hydrogen atom from the propyl or ethyl group on the other. The elimination of hydrogen leads to m/z 86 ions. The peak at m/z 86 is not abundant in the corresponding mass spectrum because a loss in hydrogen is thermodynamically unfavorable; the hydrogen radical is far less stable than the methyl or ethyl radicals (refer to the reminder about thermochemistry at the beginning of this chapter). In most cases of this kind, the elimination of a hydrogen atom via an α cleavage is not observed because it is so energetically unfavorable. The fact that four hydrogen atoms are "available" in α leads to the detection of m/z 86 ions—a sort of statistical effect. One can also note that the peak at m/z 86 apparently corresponds to two isomer ions that are impossible to differentiate in the mass spectrum.

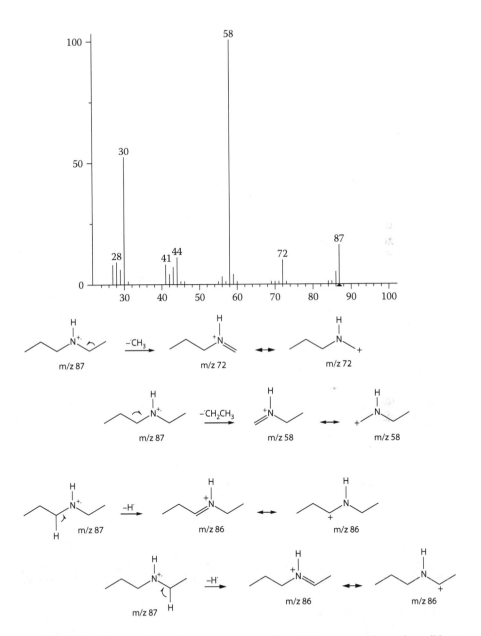

FIGURE 9.21 Electron ionization mass spectrum of n-ethyl,1-propanamine and possible α cleavages from the molecular ion.

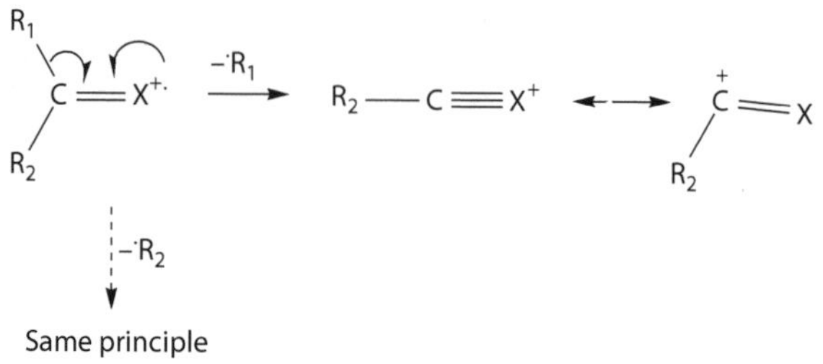

Same principle

FIGURE 9.22 Mechanism of α cleavage where the heteroatom (X) is engaged in a double bond.

Consider the case of a heteroatom carrying at least one free electron pair engaged in a double bond (oxygen atom of a carbonyl group, for example). The α cleavage is then written as indicated in Figure 9.22. Figure 9.23 presents the electron ionization spectrum of 3-hexanone and the two α cleavages possible from its molecular ion.

9.3.1.2 Allylic Cleavage

The allylic cleavage following the ionization of a double bond is a particular case of α cleavage. The ionization of a double bond corresponds to the detachment of one of the two electrons that constitute the π bond. Figure 9.24 presents the mechanism of allylic cleavage. Electron ionization leads to two mesomeric structures. On the left one, the radical is on the side of the substituent R_2. On the right one, it is on the side of the substituent R_1.

In reality, these are two limit forms of one and the same structure; the remaining electron is delocalized between the two atoms of carbon initially involved in the π bond. Writing these two structures demonstrates why two allylic cleavages are possible. The motor of each cleavage is the pairing of two electrons in α of the positive charge to lead to a π bond conjugated with the charge. The resulting cations are therefore stabilized by mesomeric effect.

Benzylic cleavage is a particular kind of allylic cleavage. As its name implies, it takes place on an aromatic structure of a C_6H_5-CH_2-R type molecule. Ionization on an aromatic ring can occur easily because π electrons are easy to remove and also because the process leads to a cation stabilized by conjugation. Figure 9.25 presents the mechanism of benzylic cleavage. The mesomeric structure represented for the initial radical cation is the most stable because it includes a tertiary radical.

As in the case of classic allylic cleavage, the pairing of electrons leads to a double bond in α of the positive charge that stabilizes the carbocation by mesomeric effect. In the particular case of benzylic cleavage, the cation rearranges itself spontaneously into a ring structure with seven carbon atoms (Figure 9.25) called the tropylium ion. This m/z 91 ion is omnipresent in benzylic compound spectra due to its high stability. It is indeed possible to write seven mesomeric forms of this carbocation assuming that the charge is equally distributed on the seven carbon atoms. Figure 9.26 presents the EI spectrum of butylbenzene and the benzylic cleavage leading to the formation of the tropylium ion from the molecular ion.

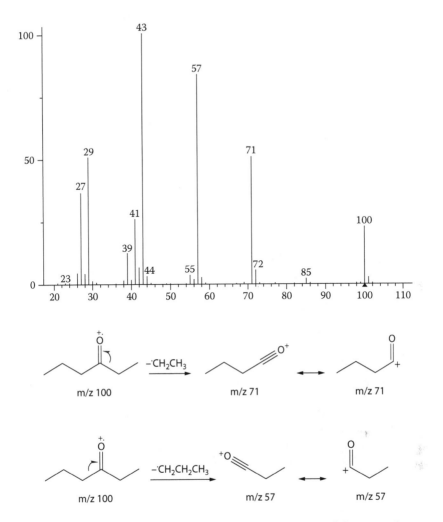

FIGURE 9.23 Electron ionization mass spectrum of 3-hexanone and the two α cleavages possible from the molecular ion.

9.3.1.3 Sigma Cleavage

Unlike the α cleavage, the sigma (σ) cleavage is a heterolytic bond cleavage. The two electrons composing the bond are captured by a single atom. This type of fragmentation is observed when the difference in electronegativity (Section 9.2.1) between the two atoms involved in the bond is very important. Usually the σ cleavages are observed from ions composed of a bond between a carbon atom and a halogen (F, Cl, Br, I).

The halogen atom is far more electronegative than the carbon atom and thus attracts both electrons of the bond, which is therefore strongly polarized. The σ

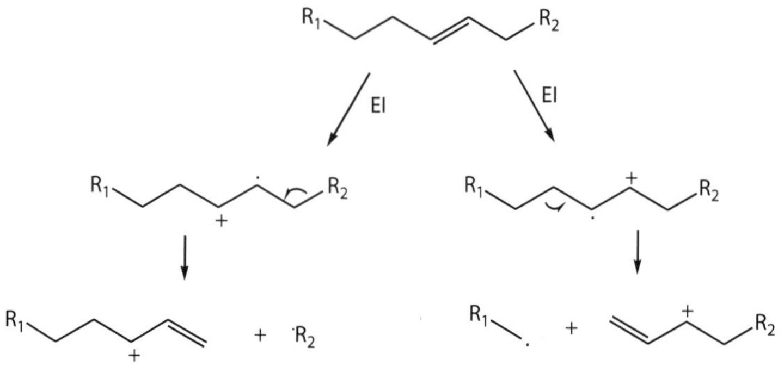

FIGURE 9.24 Mechanism of allylic cleavage following ionization of a double bond.

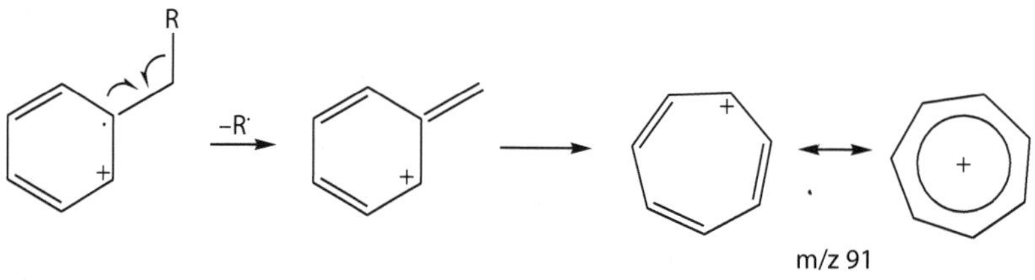

FIGURE 9.25 Mechanism of α cleavage from an ionized benzenic compound.

cleavage mechanism shown in Figure 9.27 is written with a double arrow since two electrons are involved. As with α cleavage, σ cleavage eliminates a radical from the $M^{+\bullet}$ ion and therefore leads to an ion with an even number of electrons.

Figure 9.28 presents the electron ionization spectrum of 2-bromopentane and the σ cleavage leading to the elimination of a bromine atom from the molecular ion. From an energetic view, this fragmentation requires a very small amount of internal energy (no transition stage requiring a distortion of the ion geometry), to such an extent that the totally fragmented molecular ion does not appear in the mass spectrum.

9.3.1.4 Alkanes

By definition alkanes possess no heteroatoms (no free electron pairs) or double bonds (no π electrons). They consist of simple C-C and C-H bonds and therefore count only σ electrons that are more difficult to remove than π and n electrons. The molecular ions are generally scarce because they are very unstable. The EI removal of one of the two electrons composing a σ bond leads to directly bond cleavage. Figure 9.29 shows the main ions thereby formed in the case of n-heptane. The abundance of ions decreases with the increase of the m/z ratio beyond m/z 57.

Figure 9.30 compares the EI mass spectra of decane and dodecane. The major ions are identical. One can distinguish the two molecules via the presence (in weak amounts) of the molecular ion in each of the spectra but the molecular ion is no longer visible on long chain alkanes and these are not generally differentiable by EI.

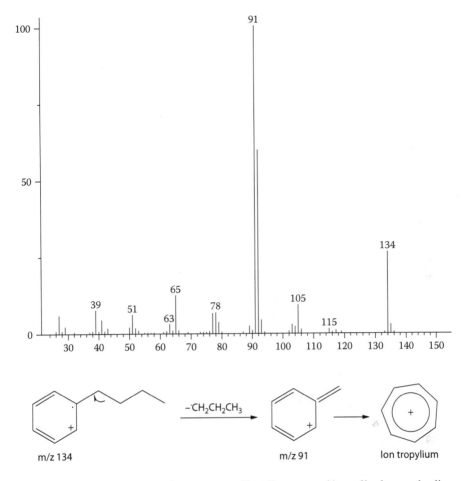

FIGURE 9.26 Electron ionization spectrum of butylbenzene and benzylic cleavage leading to formation of the tropylium ion from the molecular ion.

$$R\text{-}X^{+\cdot} \longrightarrow R^+ + X^{\cdot}$$

FIGURE 9.27 Mechanism of σ cleavage.

9.3.2 REARRANGEMENTS

The two main rearrangements observed in EI are the McLafferty rearrangement and the reaction known as the retro Diels-Alder rearrangement. The ions issued from these rearrangements are easy to identify in mass spectra based on the nitrogen rule (Section 9.6.3).

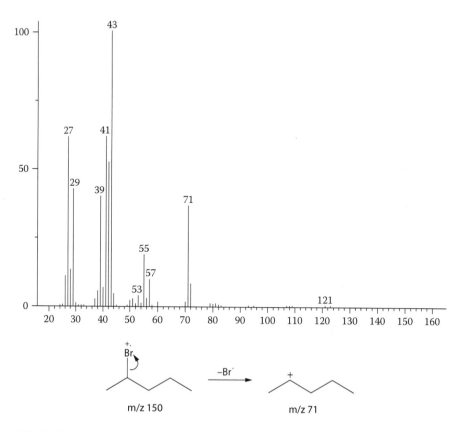

FIGURE 9.28   Electron ionization mass spectrum of 2-bromopentane and σ cleavage leading to elimination of bromine atom from the molecular ion.

9.3.2.1   McLafferty Rearrangement

The McLafferty rearrangement is presented in Figure 9.31. It is a concerted mechanism for the transfer of a hydrogen atom via a six-center transition state. As in solution chemistry, the six atom rings are particularly stable in the gas phase. The reaction therefore requires weak activation energy (see Section 9.2 covering thermochemistry parameters), which explains why the ions issued from a McLafferty rearrangement are generally abundant in mass spectra.

In Figure 9.31, there is no precision concerning which product of the reaction carries the charge because it has been demonstrated that this depends on the nature of the substituents X, Y, R_1, and R_2. Unlike the simple α and σ cleavages, a rearrangement does not induce a reversal in parity, that is, the dissociation of the $M^{+\bullet}$ ion leads to the formation of another radical ion. The product with the lowest ionization potential (generally the one at left in the figure) carries the charge and the radical at the end of the reaction. Figure 9.32 presents the EI mass spectrum of methyl 2-methylpentanoate and the McLafferty rearrangement mechanism that occurs from the molecular ion to supply the m/z 88 ion.

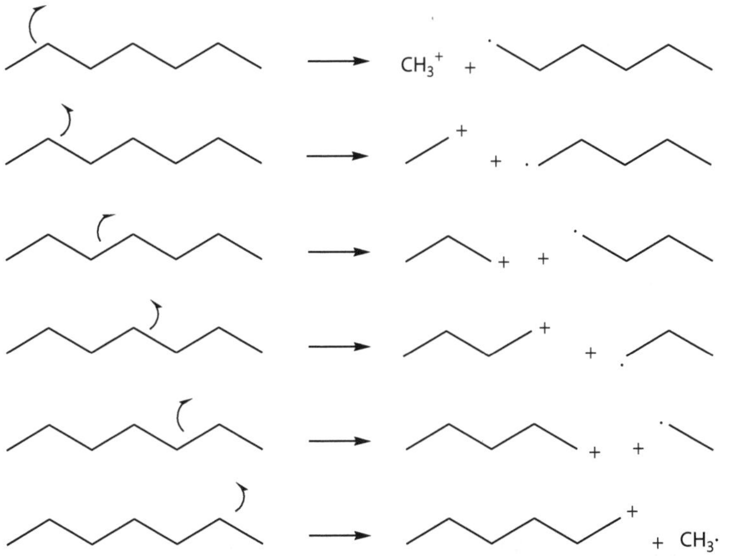

FIGURE 9.29   Fragmentations of n-heptane in electron ionization.

Another example of the McLafferty rearrangement is given for ionized butylben-zene, whose mass spectrum is displayed in Figure 9.26. In this case, the rearrange-ment leads to the m/z 92 ion according to the mechanism presented in Figure 9.33. To illustrate the mechanism, there is no need to specify the positions of the charge and radical because of the stabilization of the m/z 92 ion by the delocalization of electrons on the aromatic ring by mesomeric effect.

9.3.2.2   McLafferty Rearrangement with Double Hydrogen Transfer

A variation of the McLafferty rearrangement is a mechanism that also starts by a hydrogen atom transfer via a six-center concerted mechanism and continues with a second hydrogen transfer as indicated in Figure 9.34. This rearrangement is dif-ficult to spot in a spectrum because the transfer of two hydrogen atoms induces the elimination of a radical to lead to an ion with an even number of electrons. Therefore, there is no inversion in parity (unlike the McLafferty rearrangement) and the nitro-gen rule does not allow diagnosing this type of rearrangement.

Figure 9.35 presents the EI mass spectrum of butyl propanoate and the McLafferty rearrangement with double transfer of hydrogen atoms from the molecular ion.

9.3.2.3   Retro Diels-Alder Reaction

As its name implies, the retro Diels-Alder reaction is the opposite of the Diels-Alder reaction. It is used widely in organic chemistry for the synthesis of six carbon atom rings from an alkene and a conjugated diene. An ionized cyclohexene has the chemi-cal structure necessary for the observation of this reaction.

The retro Diels-Alder reaction mechanism is described in Figure 9.36. As for the McLafferty rearrangement, the charge and the radical are carried by the

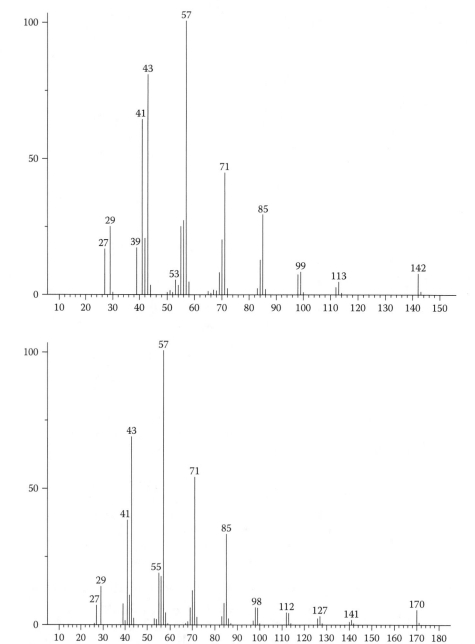

FIGURE 9.30 Electron ionization mass spectra of decane (top) and dodecane (bottom).

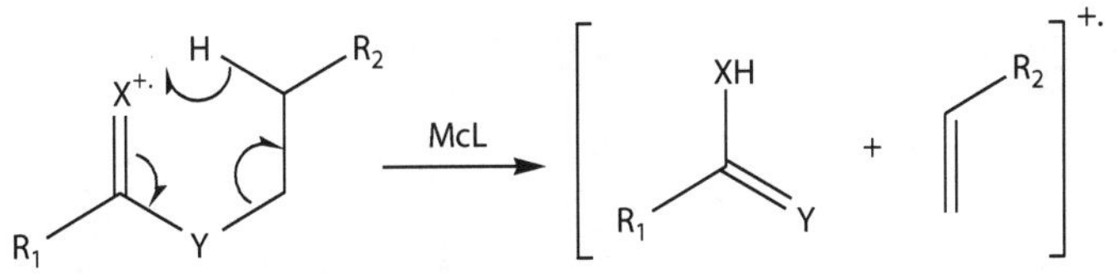

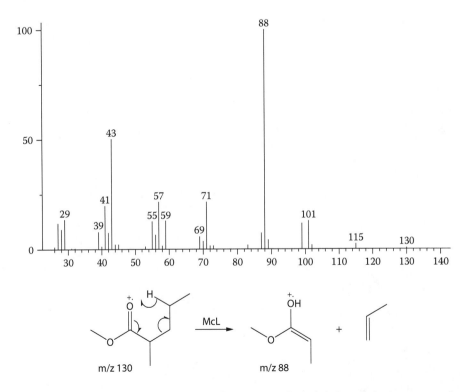

FIGURE 9.31 Mechanism of McLafferty rearrangement; X = O, S, NH, NR, CH$_2$, CHR. Y = O, S, CH$_2$, CHR, CRR′, NH, NR.

FIGURE 9.32 Electron ionization mass spectrum of methyl 2-methylpentanoate and McLafferty rearrangement occurring from the molecular ion leading to ion at m/z 88.

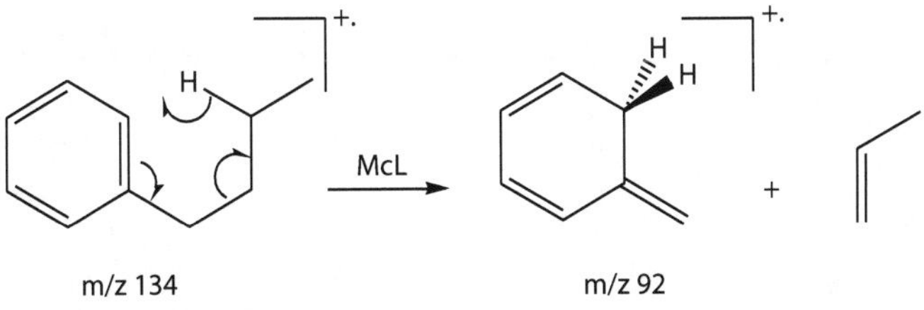

FIGURE 9.33 McLafferty rearrangement leading to ion at m/z 92 from the molecular ion of butylbenzene.

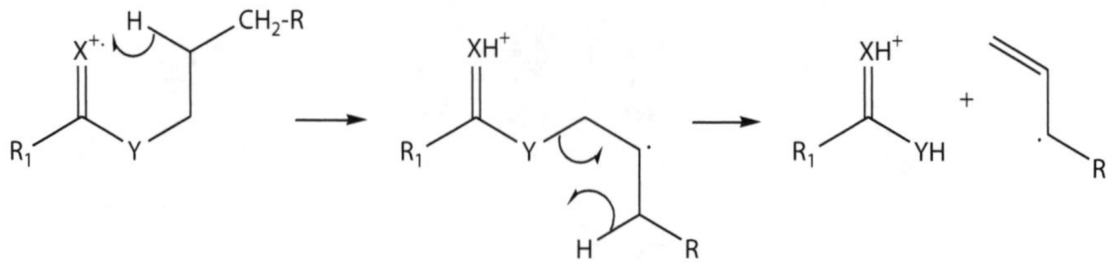

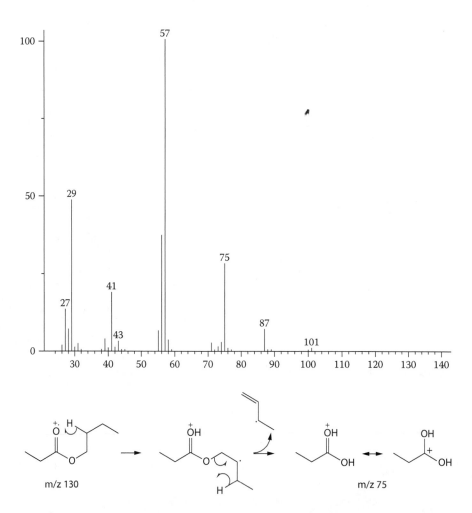

FIGURE 9.34 McLafferty rearrangement with double transfer of hydrogen atoms. X = O, S, NH, NR, CH₂, CHR. Y = O, S, CH₂, CHR, CRR′, NH, NR.

FIGURE 9.35 Electron ionization mass spectrum of butyl propanoate and McLafferty rearrangement with double transfer of hydrogen atoms from the molecular ion.

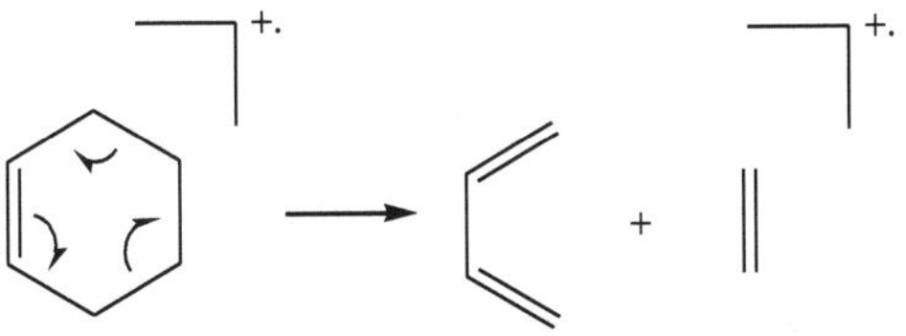

FIGURE 9.36 Retro Diels-Alder reaction.

same product—the one that has the lowest ionization potential (usually the diene). Figure 9.37 presents the electron ionization mass spectrum of 1,2,3,4-tetrahydronaphthalene and the retro Diels-Alder reaction from its molecular ion.

9.3.3 SECONDARY FRAGMENTATION OF EVEN-ELECTRON IONS

As explained previously, even-electron ions formed in EI are issued from simple α and σ cleavages. We will see in this section how these ions are susceptible to fragment in turn. After a simple cleavage, the radical is eliminated and the fragmentation of the ion is generally induced by the charge. There are nevertheless some fragmentations (charge remote) that are induced only by the internal energy of the ion and not by the charge.

9.3.3.1 Secondary Fragmentation Involving Heterolytic Cleavage

It is common for the positive charge of an even-electron ion to induce the heterolytic cleavage of a σ bond. In Figure 9.38, the ion issued from the loss of F$^•$ by σ cleavage dissociates itself to eliminate pentene by heterolytic cleavage. Figure 9.39 presents another example of fragmentation resulting from a heterolytic cleavage of a σ bond induced by a positive charge. As is often the case of carbonyl compounds, this fragmentation is followed by the elimination of carbon monoxide.

9.3.3.2 Concerted Eliminations Following Simple Cleavage

Another common type of fragmentation is elimination referred to as alpha, beta (α,β). Two atoms of atom groups carried by adjacent carbon atoms are eliminated to form a double bond conjugated with a positive charge. The resulting cation is stabilized by mesomeric effect. It is a concerted mechanism involving a four-center transition state. Figure 9.40 describes this type of fragmentation from the m/z 71 ion issued from the σ elimination of a bromine atom from 2-bromopentane (see Figure 9.28). Note that the m/z 55 ion resulting from the elimination of methane is relatively abundant in the spectrum of 2-bromopentane. In contrast, the m/z 69 ion resulting from the loss of a hydrogen molecule is observed only at trace level. This illustrates that a "chemically" possible fragmentation does not systematically occur in the source of the mass spectrometer.

Another type of mechanism that is just as common is the elimination of an alkane with the concerted transfer of a hydrogen onto the carbon skeleton. Figure 9.41 presents this mechanism for the previously mentioned ion at m/z 71 that loses an ethylene molecule to lead to the m/z 43 ion that is very abundant in the spectrum of 2-bromopentane (Figure 9.28).

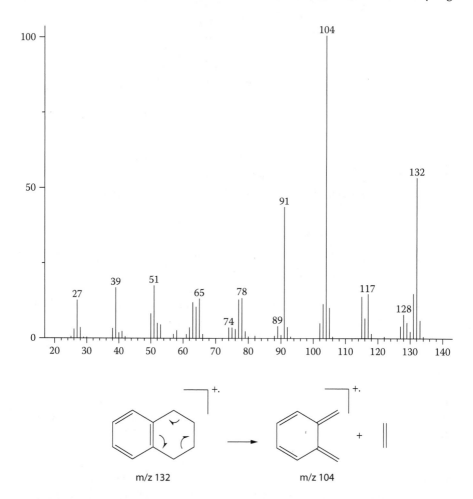

FIGURE 9.37   Electron ionization mass spectrum of 1,2,3,4-tetrahydronaphthalene and retro Diels-Alder reaction from the molecular ion.

FIGURE 9.38   The σ cleavage followed by charge-induced elimination of pentene.

Why is this type of mechanism so common even though it is not directly induced by the charge? The reason is the particular nature of transfers involving four centers. Consider the mechanism detailed in Figure 9.42. Four electrons represented by the black points are involved. The mechanism can be written in three ways (a, b, and c in the figure). Mechanism a is written with simple arrows. One electron of the C_1-C_2 bond pairs with one of the electrons of the C_3-H bond to form the C_1-H bond. The two remaining electrons pair to form the C_2-C_3 double bond. In mechanism b, the C_1-H

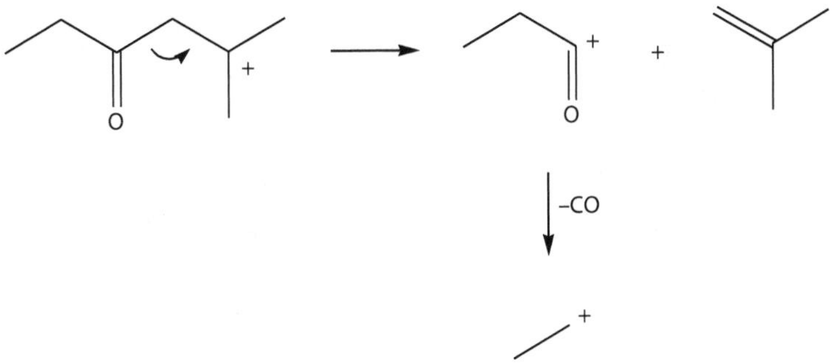

FIGURE 9.39 Charge-induced heterolytic cleavage of a σ bond and consecutive elimination of carbon monoxide.

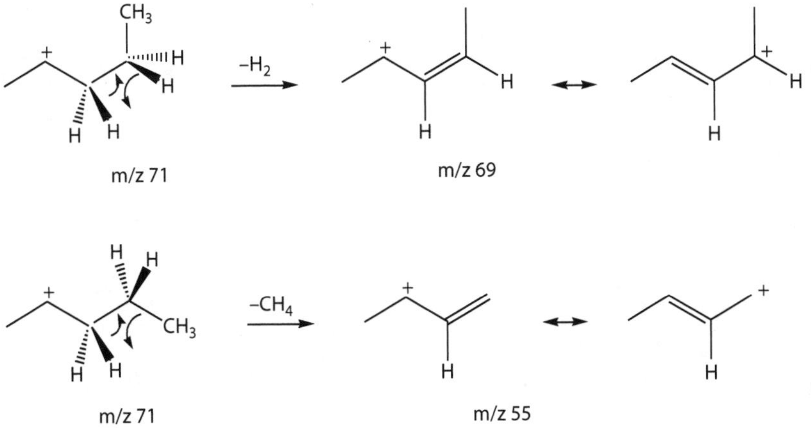

FIGURE 9.40 Eliminations of dihydrogen (top) and methane (bottom) from a m/z 71 ion.

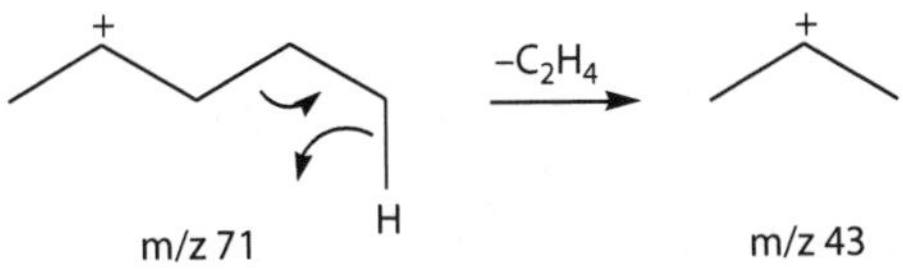

FIGURE 9.41 Ethylene elimination from m/z 71 ion.

bond is formed by the two electrons initially constituting the C_3-H bond whereas the two electrons of the C_1-C_2 bond move to form the C_2-C_3 π bond. Mechanism c is the opposite of mechanism b.

The reactions b and c are written with double arrows because they involve transfers of electron pairs, not single electrons. The three depictions of the elimination presented in Figure 9.42 lead to the same final state. Choosing among them is simply a question of formalism. These are three correct ways of writing the same mechanism.

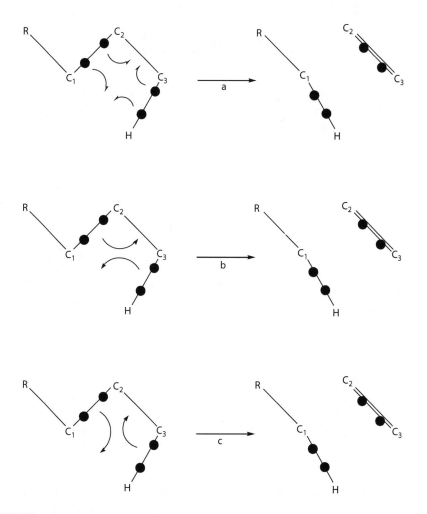

FIGURE 9.42   Three ways of writing elimination via four-membered transition state.

Indeed, the structure must only vibrate sufficiently for the temporary decrease of the $(C_1\text{-}C_2\text{-}C_3)$ and $(C_2\text{-}C_3\text{-}H)$ angles to allow the reorganization of electrons and the corresponding fragmentation. It is generally acceptable to adopt one of the double arrow versions (b or c) because the mechanism implies an even-electron ion at the start.

9.3.3.3   Carbon Monoxide Elimination after Alpha Cleavage

The α cleavage induced by the ionization of a carbonyl function (ketone, lactone, ester, etc.) is almost always accompanied by the elimination of carbon monoxide (CO) according to the mechanism in Figure 9.43. This fragmentation is easy; it requires very little internal energy because the activation energy of the reaction is very weak. The loss of CO corresponds to the "infinite" expansion of the C-CO bond; the reaction does not require passing through a specific transition state.

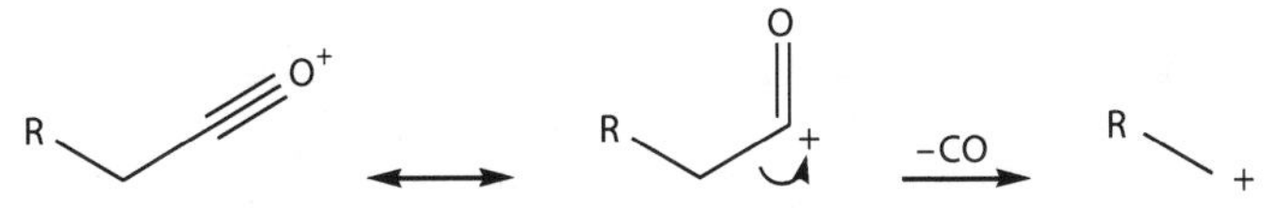

FIGURE 9.43 Carbon monoxide elimination following cleavage in α of a carbonyl group.

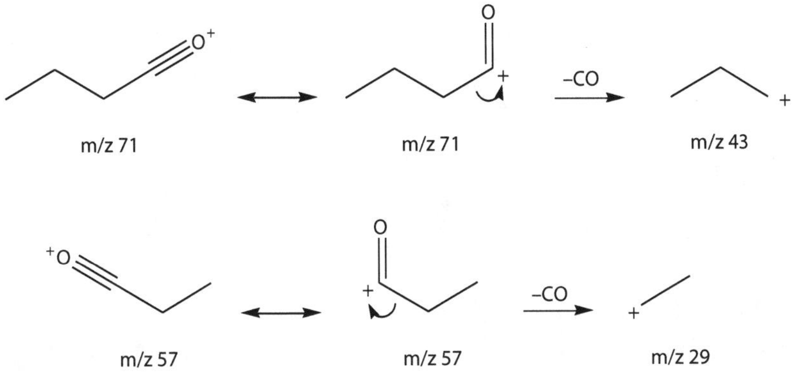

FIGURE 9.44 Elimination of carbon monoxide from m/z 71 and m/z 57 ions issued from ionized 3-hexanone.

Figure 9.44 depicts the elimination of carbon monoxide from m/z 71 and m/z 57 ions issued from ionized 3-hexanone (see the Figure 9.23 spectrum). The resulting m/z 43 and m/z 29 ions are abundant in the corresponding mass spectrum.

Esters represent a specific case, as shown in Figure 9.45. When the α cleavage takes place on the side of the alcohol function (α1), the $R_2O^•$ radical is eliminated and the resulting carbocation undergoes the elimination of carbon monoxide. When the α cleavage takes place on the side of the acid function (α2), the R_1CH_2. radical is eliminated. The fragmentation stops at this stage because the oxygen atom of the alcohol function of ester is too electronegative to allow the departure of CO with the two electrons of the O-CO bond. The elimination of CO from the resulting cation

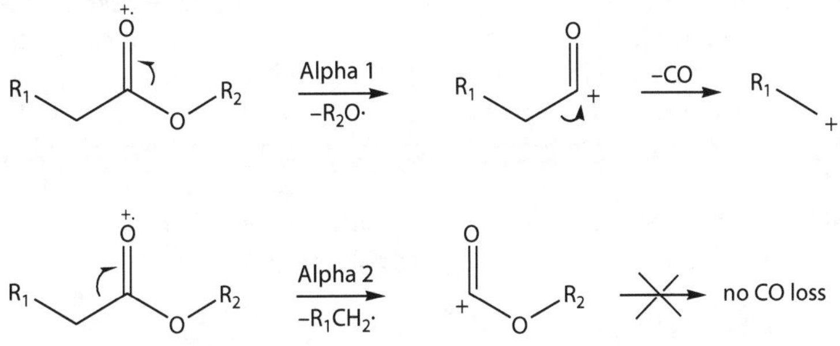

FIGURE 9.45 The α cleavages and CO elimination of esters.

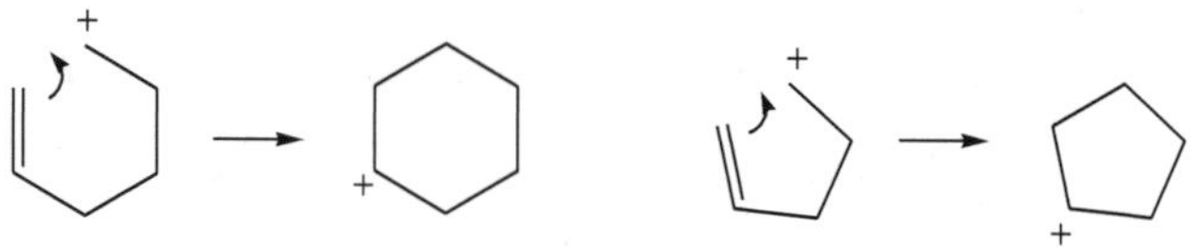

FIGURE 9.46   Charge-induced cyclization of alkene cations.

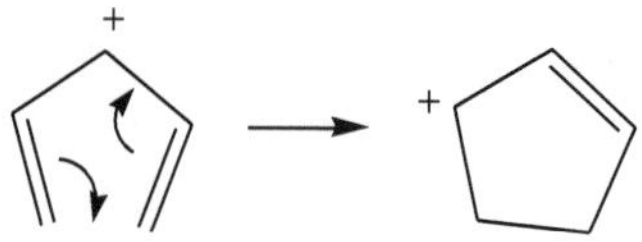

FIGURE 9.47   Nazarov mechanism.

would lead to the formation of a R_2-O^+ even-electron ion, with the positive charge carried by the oxygen atom and therefore very unstable.

9.3.3.4   Ion Cyclization

Because it diminishes the entropy of the ion (its flexibility and the number of degrees of freedom) and consequently its internal energy, the cyclization of ions is a very frequent phenomenon observed in the gas phase. For example, the formation of five or six center rings from an alkene, according to the charge-induced mechanisms described in Figure 9.46, is very frequent. In the examples suggested, this cyclization is even more favorable because it leads to a carbocation that is stabilized better by the electron-releasing inductive effect because it is disubstituted.

A cyclization mechanism of the same kind known as the Nazarov reaction is presented in Figure 9.47. The positive charge is stabilized by only one +M effect in the final state against two +M effects in the initial state but the decrease of entropy contributes to the decrease in internal energy of the system.

9.3.4   Secondary Fragmentation of Odd-Electron Ions

9.3.4.1   Principle

The secondary fragmentation of odd-electron ions is generally dominated by the reactivity of the radical; the free electron tries to pair with a neighbor electron as is the case in α cleavage. The ions issued from the opening of a six-center ring following an α cleavage fragment themselves in this exact manner. The other radical ions issued from a first fragmentation of the molecular ion result from rearrangements (e.g., McLafferty rearrangement or retro Diels-Alder reaction) or some distonic ions (odd-electron ions whose charges and radicals are carried by different atoms). Figure 9.48 presents frequent dissociation mechanisms of distonic ions.

9.3.4.2   Alicyclic Compounds

The ionized alicyclic compounds present a specific behavior after α cleavage. After the ionization of the heteroatom adjacent to the cycle, the α cleavage leads to the opening of the cycle. Consider the ionized aniline presented in Figure 9.49. The ion

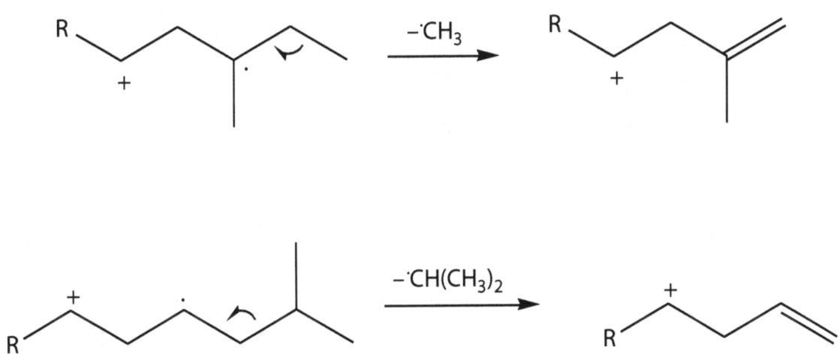

FIGURE 9.48 Examples of radical-induced dissociations of distonic ions.

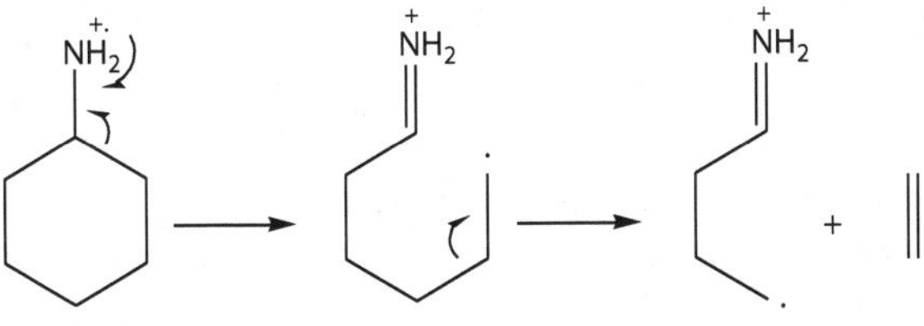

FIGURE 9.49 The α cleavage from ionized aniline.

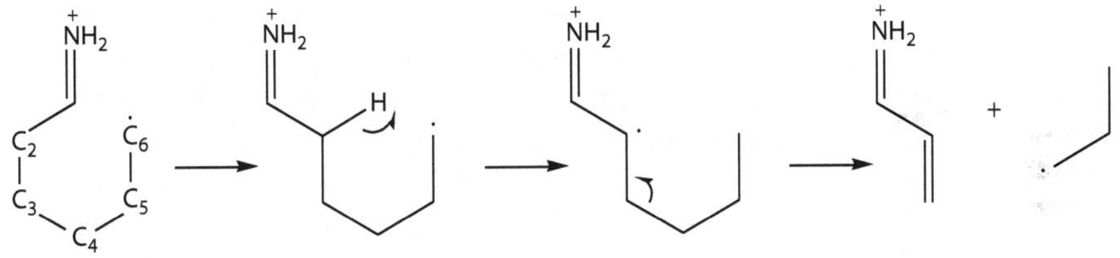

FIGURE 9.50 Evolution of ion B issued from α cleavage from ionized aniline (see Figure 9.49).

B issued from this cleavage possesses the same m/z ratio as the initial ion A since no radical was eliminated at this stage. The loss of ethylene leading to structure C from ion B is possible but not often observed since B rearranges itself spontaneously as indicated in Figure 9.50.

Ion B no longer has the fixed structure of ionized aniline. The opening of the cycle allows free rotation around the C_2-C_3 bond to lead to structure B′ presented in Figure 9.50. B′ corresponds to a transition stage involving six centers, therefore requiring low internal energy and allowing the transfer of a hydrogen atom from the carbon atom C_2 to the carbon atom C_6. The driving force of the reaction is the stabilization of the radical since the transfer of the hydrogen atom transforms a primary radical (structure B) into a secondary and conjugated radical (structure D). The fragmentation D → E corresponds to the pairing of radicals leading to a π bond that contributes to the stabilization of cation E by mesomeric effect.

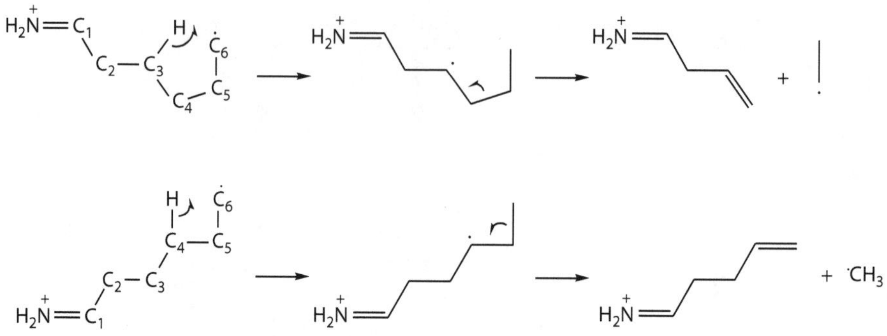

FIGURE 9.51 Hydrogen atom transfers through five-membered (top) and four-membered (bottom) transition states and consecutive radical-induced fragmentations.

The mechanism in Figure 9.50 is favored due to the peculiar stability of six-center structures in chemistry. It can nevertheless be in competition with the transfers of hydrogen atoms carried by carbons C_3 and C_4. Figure 9.51 illustrates these types of transfers via more strained transition states involving five- and four-membered structures.

The mechanisms presented in Figure 9.51 are energetically less favorable than the mechanism presented in Figure 9.50. One can therefore expect the corresponding ions to be weakly abundant in the spectra. The presence of an alkyl substitute in the cycle can nevertheless modify this logic. As an example, Figure 9.52 shows how the presence of an isopropyl substitute in C_3 induces an efficient stabilization of the radical issued from hydrogen transfer via a five-membered structure. The hydrogen transfer in this case allows transforming a primary radical into a tertiary radical. Figure 9.53 presents the mass spectrum of cyclohexanamine and the fragmentations of the molecular ion leading to m/z 56 and m/z 70 ions.

The two mechanisms following an α cleavage are the eliminations of a propyl radical and an ethyl radical. The first implies a six-membered transition state requiring

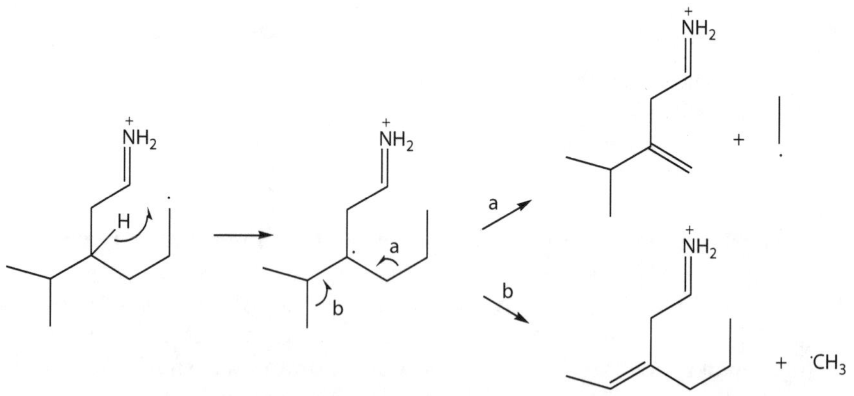

FIGURE 9.52 Hydrogen transfer through five-membered transition state following α cleavage and consecutive radical-induced fragmentations.

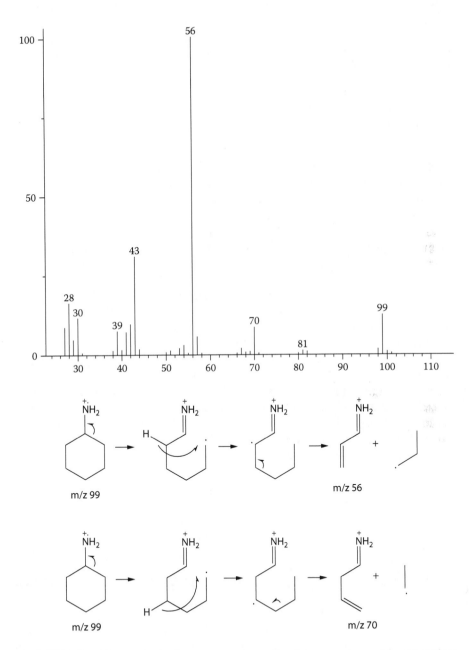

FIGURE 9.53 Electron ionization mass spectrum of cyclohexanamine and fragmentations of the molecular ion leading to m/z 56 and m/z 70 ions.

little energy; the resulting ion therefore constitutes the base peak of the spectrum. The ethyl radical loss requires passing through a more strained five-membered transition state; that is why the m/z 70 ion is less abundant in the spectrum than the m/z 56 ion. One can notice the absence of the m/z 71 ion that would result from direct elimination of ethylene after the opening of the ring. This illustrates the priority character of the hydrogen transfer on the primary radical directly issued from α cleavage.

9.4 DISSOCIATION MECHANISMS IN POSITIVE CHEMICAL IONIZATION

The ion formation mechanisms in positive chemical ionization are presented in Chapter 3. The $M^{+\bullet}$ ions issued from charge exchange almost do not fragment in the conditions under which they are formed in GC-MS (unlike the $M^{+\bullet}$ ions formed by electron ionization). Methane chemical ionization often directly supplies fragment ions by removing a hydride or alkyl anion.

The adduct ions such as MNH_4^+ do not fragment significantly in the mass spectrometer or dissociate themselves into $M + NH_4^+$ at m/z 18 in case of an increase in internal energy (in MS/MS in particular)—which is of absolutely no interest from an analytical view.

We are interested exclusively in the fragmentation of protonated molecules. Compared to EI, CI is a "soft" ionization mode. This means that the MH^+ ions issued from CI have much less internal energy than the $M^{+\bullet}$ ions produced in EI and therefore fragment themselves much less than the latter.

Unlike the main fragmentations in electron ionization induced by the radical, the fragmentations observed in positive CI are mostly induced by the charge. As Figure 9.54 displays, the protonation of an NH_2 or OH group will generally be accompanied by the elimination of NH_3 or H_2O; the heteroatom is attempting to recover its free electron pair. Another kind of dissociation observed in positive chemical ionization is accompanied by the transfer of a hydrogen atom on the protonated site, as depicted in Figure 9.55 for butylbenzene.

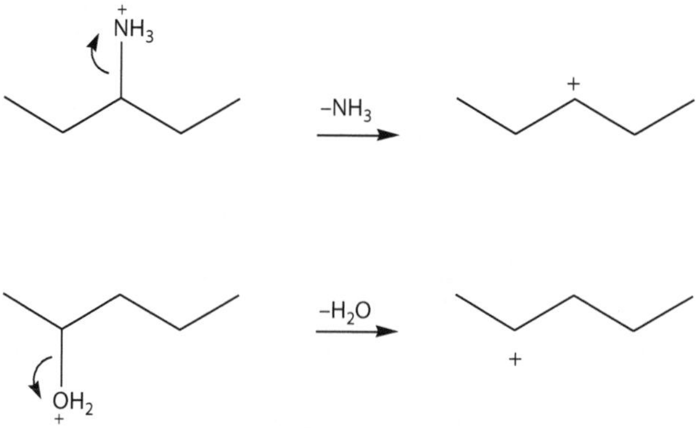

FIGURE 9.54 Elimination of ammonia from protonated primary amine (top) and water from protonated alcohol (bottom) in positive chemical ionization.

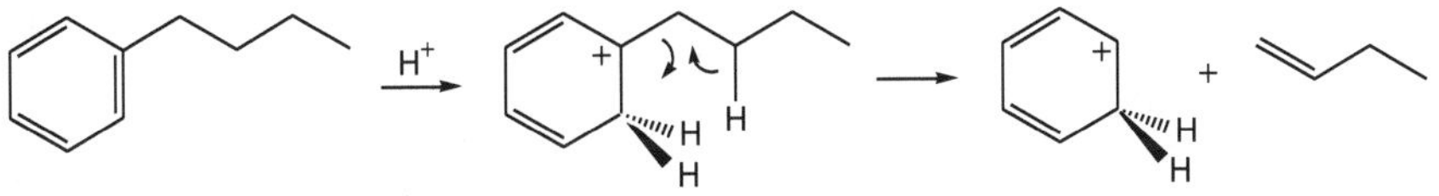

FIGURE 9.55 Protonation of butylbenzene followed by hydrogen transfer on the charged carbon atom leading to butene elimination in positive chemical ionization.

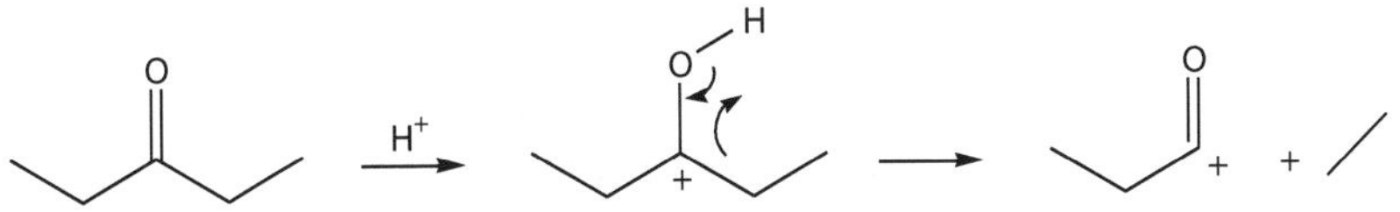

FIGURE 9.56 Protonation of 3-pentanone and ethane elimination after hydrogen transfer on the alkyl chain in positive chemical ionization.

By an analogous mechanism, the proton issued from the reactant can be transferred from the ionization site onto the carbon skeleton, in the case of a ketone, for example. This mechanism is described for the case of 3-pentanone in Figure 9.56. The other fragmentations observed in positive chemical ionization are related to those described in Section 9.3.3: fragmentations induced by the charge (concerted α,β eliminations, for example) or charge remote dissociations. One can consequently refer to this section to interpret most mass spectra recorded in positive CI.

9.5 DISSOCIATION MECHANISMS IN NEGATIVE CHEMICAL IONIZATION

Negative ionization remains an ambiguous term in GC-MS coupling. In reality, negative chemical ionization has two different ways of operating. The only point they have in common is that they supply negative ions. The most used negative chemical ionization technique in the scope of GC-MS coupling analyses is electron attachment (see Chapter 3). It is also possible to produce ions by deprotonation but this process is more complex to perform on most commercial mass spectrometers. The mechanisms involved in the two techniques are described below.

9.5.1 DISSOCIATION OF IONS PRODUCED BY ELECTRON ATTACHMENT

As noted above, the ions formed by electron attachment are negative radical (odd-electron) ions. We have previously seen how much the $M^{+\bullet}$ radical ions are reactive in electron ionization. The $M^{-\bullet}$ radical ions are generally less so, mainly because the ionizing electrons are thermalized in electron attachment and consequently the internal energy of the resulting ions is weak.

The spectra are therefore often dominated by a molecular ion of m/z ratio M, which by the way constitutes the main objective of electron attachment users. This is why the dissociation of $M^{-\bullet}$ ions has been the subject of very few studies compared to $M^{+\bullet}$ and MH^{+} ions. Figure 9.57 compares the electron ionization and electron

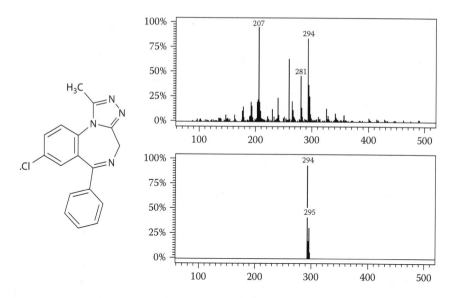

FIGURE 9.57 Mass spectra of alprazolam recorded in electron ionization (top) and in electron attachment (bottom).

attachment mass spectra of alprazolam (benzodiazepine). One can clearly see the difference in fragmentation ratios according to the considered ionization mode. The daughter ions in EI are numerous and are absent in electron attachment.

Most applications of electron attachment concern polyhalogenated aromatic compounds, particularly polychlorobenzenes (PCBs), polybromodiphenylethers (PBDEs), and dioxins.[9] An electron will attach itself naturally to the molecule on the most electrophilic site. When a halogen is present, the negative charge and the radical will be localized on the halogen. The problem is that the electronegativity of the halogen often leads to its departure with the charge, as indicated in Figure 9.58 in the case of a congener of dibromodiphenylether. It is generally considered that the departure of Br⁻ is simultaneous to the electron attachment; the process is called dissociative capture.

The problem for the chemist is that the Br ions at m/z 79 and m/z 81 (due to the two bromine isotopes) are absolutely not specific of the molecule and inform neither on the nature of the molecule nor its bromination degree. The same kind of problem is encountered with PCBs that mainly supply Cl⁻ chlorine ions at m/z 35 and m/z 37

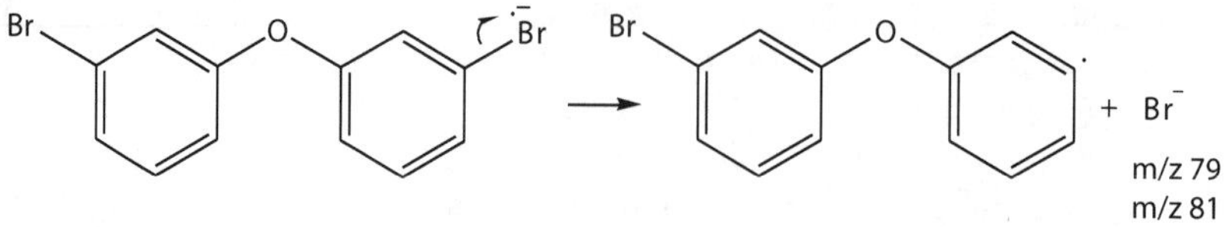

FIGURE 9.58 Loss of Br⁻ from a congener of dibromodiphenylether ionized by electron attachment.

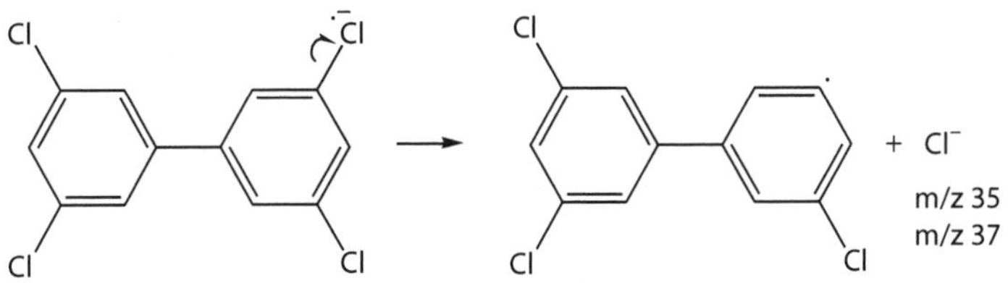

FIGURE 9.59 Loss of Cl⁻ from 3,3′,5,5′-tetrachlorobiphenyl under electron attachment.

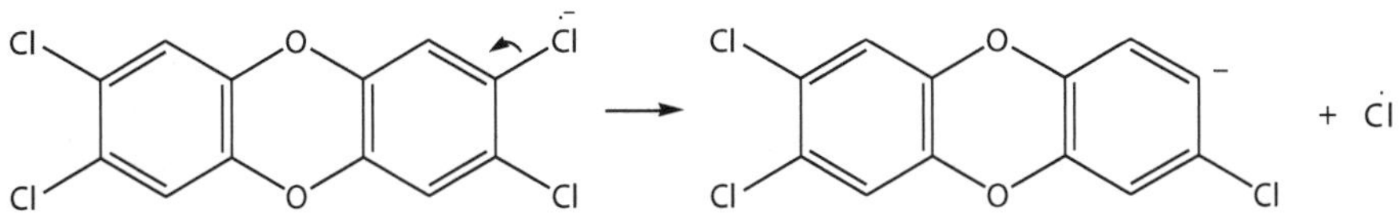

FIGURE 9.60 Formation of [M-Cl]⁻ ion from tetrachlorodioxin under electron attachment.

and molecular ions that are weakly abundant, in particular for less chlorinated congeners.[10] An example of Cl⁻ loss from 3,3′,5,5′- tetrachlorobiphenyl is presented in Figure 9.59.

The loss of the X⁻ halide is in competition with the corresponding X· radical that supplies the [M-X]⁻ ion. This last mechanism is presented in Figure 9.60 in the case of a tetrachlorodioxin. Unlike the elimination of halide ions that results from homolytic cleavage of the C-X bond, the formation of [M-X]⁻ results from a heterolytic cleavage. For polyhalogenated aromatic compounds, the relative proportions of $M^{·-}$, [M-X]⁻, and X⁻ in the mass spectrum depend on the conditions of the spectrometer source (temperature, pressure) and also and especially on the nature of the thermalization gas used.[11]

Polychloroalkanes constitute a class of molecules that are studied frequently in electron attachment. These molecules are generally formed by the chlorination of alkenes and designated Mx,y; in this notation, x and y, respectively, indicate the numbers of carbon and chlorine atoms. The ions observed in mass spectra obtained by electron attachment are $Cl_2^{·-}$, HCl_2^-, [M – Cl]⁻, [M – HCl]⁻·, [M – HC – Cl]⁻, [M – 2HCl]⁻·, [M – HCl – Cl_2]⁻, [M – Cl_2 – Cl]⁻, and [M + Cl]⁻. The relative proportions of these different ions are very dependent on the conditions of the source, particularly the temperature and pressure. The formation of the [M + Cl]⁻ ion (adduct between the molecule and the chlorine ion) is explained by the two-step mechanism described in Figure 9.61. A chlorine ion is first produced by dissociative electron capture (reaction 1); it then attaches itself to a neutral molecule (reaction 2).

$$M + \tilde{e}_{(thermalized)} \rightarrow [M - Cl]^· + Cl^- \qquad [1]$$
$$M + Cl^- \rightarrow [M + Cl]^- \qquad [2]$$

FIGURE 9.61 Two-step mechanism of formation of [M + Cl]⁻ ions displayed in electron attachment mass spectra of chloroalkanes.

The examples described previously demonstrate that the mass spectra interpretations of the most frequently analyzed molecules in electron attachment involve no sophisticated mechanisms, unlike interpretation in EI. The elimination of a halide, with or without a negative charge, constitutes the main dissociation mechanism of $M^{\cdot-}$ ions.

9.5.2 MECHANISMS OF DEPROTONATION

Negative chemical ionization by proton removal (deprotonation) by a base to provide $(M - H)^-$ ions is used rarely now except in research laboratories. It concerns molecules whose acidic hydrogens are easy to detach with a base. If one wants to use hydroxyl OH^- ions as basic reagents, one can for example form a hydroxyl by using a mixture of nitrous oxide and methane. Figure 9.62 illustrates the mechanisms of dissociative electron capture on N_2O to form the $O^{\cdot-}$ ion (reaction 3) and abstraction of a hydrogen atom from methane by $O^{\cdot-}$ to form OH^- (reaction 4). The OH^- ion can then abstract a proton from the analyte to form the $(M - H)^-$ ion (reaction 5).

With the ion traps that offer the possibility of performing negative chemical ionization with liquid reactants, methanol is a preferred reactant in negative mode. The reagent ions are then methanolate ions CH_3O^- formed according to reactions 6 and 7 in Figure 9.63.

The CH_3O^- ions can then react with the M molecule of the analyte according to the three reactions in Figure 9.64: proton abstraction (reaction 8), charge exchange

$$N_2O + \bar{e} \rightarrow N_2 + O^{\cdot-} \qquad [3]$$

$$CH_4 + O^{\cdot-} \rightarrow OH^- + CH_3. \qquad [4]$$

$$OH^- + M \rightarrow (M\text{-}H)^- + H_2O \qquad [5]$$

FIGURE 9.62 Reactions leading to deprotonation of analyte M in negative chemical ionization using mixture of nitrous oxide and methane.

$$CH_3OH + \bar{e}(70\ eV) \rightarrow (CH_3OH)^{-\cdot*} \qquad [6]$$

$$(CH_3OH)^{-\cdot*} + CH_3OH \rightarrow CH_3O^{\cdot} + CH_3O^- + H_2 \qquad [7]$$

FIGURE 9.63 Formation of methanolate CH_3O^- ions from methanol in negative chemical ionization.

$$CH_3O^- + M \rightarrow CH_3OH + [M\text{-}H]^- \qquad [8]$$

$$CH_3O^- + MX \rightarrow MX^{-\cdot} + CH_3O^{\cdot} \qquad [9]$$

$$CH_3O^{\cdot} + MX \rightarrow CH_3OX + M^- \qquad [10]$$

FIGURE 9.64 Possible reactions of CH_3O^- ions with analyte M.

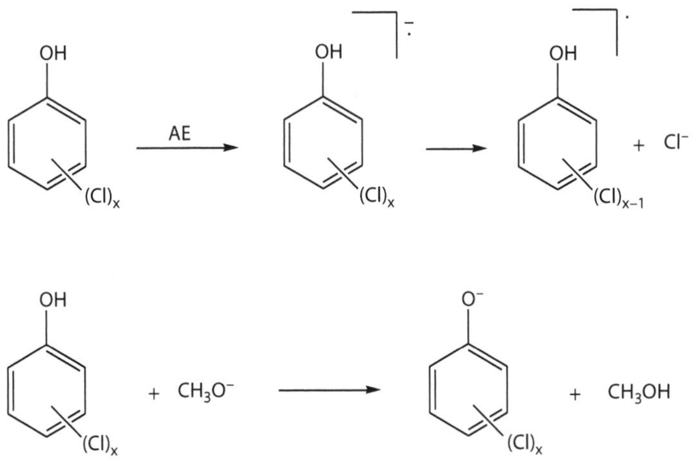

FIGURE 9.65 Ionization of chlorophenol by electron attachment (top) and by proton abstraction (methanol negative chemical ionization, bottom).

(reaction 9), or nucleophilic substitution (reaction 10). The reaction pathway depends on the nature of the analyte.

One of the applications of proton abstraction is illustrated in Figure 9.65, which shows that a chlorophenol provides under electron attachment an $M^{\cdot-}$ ion that fragments itself to yield a Cl^- ion; the latter is not specific to the molecule and does not allow characterizing the corresponding chlorophenol, whereas proton abstraction supplies a $(M-H)^-$ ion at $m/z = M-1$ that is more stable than $M^{\cdot-}$ and above all specific to the molecule. As shown in this example, the point of chemical ionization by deprotonation, in comparison with electron attachment, is the absence of fragmentation of the pseudomolecular ion $(M-H)^-$ with most analytes, so there is no point in discussing dissociation mechanisms.

9.6 MASS SPECTRA INTERPRETATION STRATEGY

9.6.1 GENERAL APPROACH FOR STRUCTURE ELUCIDATION

9.6.1.1 General Points

A mass spectrum is a two-dimensional graph. The m/z ratios of ions are shown on the abscissa axis and the relative abundances are shown on the ordinate axis. Figure 9.66 shows the mass spectrum of dimethyl propanedioate in EI. The largest (base) peak of the spectrum (m/z 101 on Figure 9.66) is used as a reference for the relative abundances. The abundances of the other ions are expressed as percentages of the base peak.

In EI, the point is generally *not* to interpret all the ions present in the spectrum. The ions are so numerous that interpretation would be long and tedious. The point is to search for the structures of the most abundant and most specific ions. In the first approximation, we consider that the higher the m/z ratio of an ion, the more it is specific. If one considered, for example, the spectrum of dimethyl propanedioate

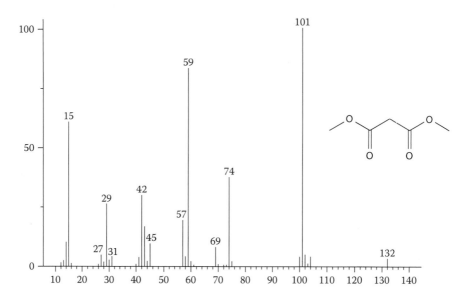

FIGURE 9.66 Electron ionization mass spectrum of dimethyl propanedioate.

presented in Figure 9.66, the interest would be in searching for the structure of ion m/z 104, despite its weak abundance, because the loss of 28 amu from the molecular ion can supply good information about its structure.

In contrast, the presence of ions at m/z 29 ($CH_3 - CH_2^+$ or $^+HC = O$) and m/z 15 (CH_3^+) is very common. They are observed on a multitude of mass spectra as soon as ion scanning over a certain m/z ratio range allows their detection. It is therefore rare for the observation of these ions to supply pertinent information about the structure of an unknown compound. Ions with low m/z ratios are often issued from consecutive fragmentations of ions of high internal energy (see Chapter 3) that do not represent statistically the global population of formed molecular ions.

9.6.1.2 Charge State of Ions

In electron ionization and in chemical ionization, monocharged ions are mostly formed, that is, z = 1 in electron ionization and positive chemical ionization and z = –1 in negative chemical ionization. Gas phase chromatography analysis is limited to small and average compounds (see Chapter 1). Beyond a certain molecular weight, the compounds are no longer volatile no matter what their polarity. In the gas phase, it is impossible for a small system to carry several charges of the same sign.

In the absence of stabilization by solvation, the electrostatic repulsion between the two charges is such that the system dissociates instantly. Only large molecules (peptides, for example) can carry several charges in the gas phase. While multicharged ions are observed rarely in GC-MS, they are omnipresent in LC-MS coupling studies of large molecules. In that case, the distances between the charges are such that the Coulomb repulsion is minimal.

9.6.1.3   Neutral Losses

After a molecular ion is identified, the first step is usually researching the structure of the ion whose m/z ratio is immediately below that of the molecular ion. Each of the ions considered pertinent (according to their relative abundance) is then analyzed by proceeding from the highest m/z ratio to the lowest. The first process consists in determining the origin of the considered ion. Can it result from the dissociation of the ion whose m/z ratio is immediately higher? (Consecutive fragmentations are very common in mass spectrometry.) Can the ion have come from a heavier ion?

We can consider the spectrum presented in Figure 9.66 as an example. Can the m/z 101 ion have come from the dissociation of the m/z 104 ion? No, because a 3 amu loss is chemically impossible. We next consider that the m/z 101 ion probably issued directly from the m/z 132 ion. Can the m/z 69 ion have come from the m/z 74 ion? No, because a 5 amu loss is also chemically impossible. Can the m/z 69 ion have issued from the m/z 101 ion? This is possible because a 32 amu loss is chemically possible (elimination of methanol). Obviously, this does not mean that the m/z 69 ion issued from m/z 101, but it represents a serious hypothesis to be investigated. Is such a transition chemically possible? To help answer this question, Table 9.3 indexes the losses of neutrals usually observed in EI. The eliminated neutral species may be either molecules or radicals.

In Table 9.3, a loss of 4 amu is not included even though consecutive eliminations of two hydrogen molecules could be considered. In the same manner, a loss of 20 amu (not shown in the table) could result from consecutive eliminations of hydrogen and water, for example. In reality, except under certain very limited conditions, observing no traces in a spectrum of a reactionary intermediate between two consecutive neutral losses is highly improbable due to strong distribution of internal energy of the ions.

TABLE 9.3
Losses of Neutrals in Electron Ionization

Mass of Neutral Eliminated (amu)	Species Eliminated	Mass of Neutral Eliminated (amu)	Species Eliminated
1	$H\cdot$	30	$CH_3\text{-}CH_3$, H_2CO
2	H_2	31	$CH_3O\cdot$
15	$CH_3\cdot$	32	CH_3OH
16	CH_4	35	$^{35}Cl\cdot$
17	NH_3, $\cdot OH$	36	$H^{35}Cl$
18	H_2O	37	$^{37}Cl\cdot$
19	$F\cdot$	38	$H^{37}Cl$
20	HF	41	$CH_2 = CH\text{-}CH_2\cdot$
26	$HC\equiv CH$, CN	42	$CH_2 = CH\text{-}CH_3$
27	$H_2C = CH\cdot$, HCN	43	$CH_3\text{-}CH_2\text{-}CH_2\cdot$
28	$CH_2 = CH_2$, CO, N_2	44	$CH_3\text{-}CH_2\text{-}CH_3$, CO_2
29	$CH_3\text{-}CH_2\cdot$, HCO		

9.6.2 MASS SPECTROMETRY AND STABLE ISOTOPES

Isotopes of a chemical element possess the same number of electrons and protons and different numbers of neutrons. For example, hydrogen and deuterium are isotopic elements. They both possess one proton and one electron. Hydrogen (^1H) has no neutron whereas deuterium (^2H or D) has one. Carbon 12 (^{12}C) has 6 electrons, 6 protons, and 6 neutrons. Carbon 13 (^{13}C) has 6 electrons, 6 protons, and 7 neutrons. The physico-chemical properties of isotopes are very similar because neutrons exert very little influence on reactivity.

The structures of many compounds include atoms in the forms of several natural isotopes. Table 4.1 in Chapter 4 lists the natures and relative abundances of the various isotopes of the main chemical elements constituting analyzable molecules in GC-MS coupling. The mass spectra present grouped peaks corresponding to the statistical distributions of the isotopes present. Ions consisting of the same atoms with different isotopes are known as isotopic isomers or isotopomers.

In the case of chemical elements presenting several isotopes in relative proportions that are sufficiently abundant for each one to be detectable, as is the case of chlorine and bromine (see below), the shape of an isotope pattern is essential for the interpretation of a mass spectrum because it reveals the number of chlorine or bromine atoms included in the raw formula of the ion. To illustrate isotope patterns, we will study carbon, chlorine, and bromine elements.

9.6.2.1 Carbon Isotopes

Carbon is naturally present in the forms of different isotopes. As indicated in Table 4.1, ^{12}C is predominant and represents 98.9% of the total. ^{14}C, commonly used for radioactive dating of ancient objects, is so scarce in nature that it cannot be detected with an ordinary mass spectrometer. ^{13}C constitutes approximately 1.1% of the total carbon and is detectable by mass spectrometry. The peak corresponding to the isotopomer of ^{13}C at m/z +1 compared to that of ^{12}C is as abundant as the number of carbon atoms of the ion.

For example, in the spectrum of pentyl 4-phenoxybenzyl succinate (raw formula $C_{22}H_{26}O_5$; Figure 9.67), we can distinguish two isotopomer $M^{+\bullet}$ ions at m/z 370 and m/z 371: the first includes only ^{12}C atoms and the second includes an atom of ^{13}C. The ^{13}C isotopomer is not visible for ions m/z 342, m/z 300, and m/z 282 because it is not sufficiently abundant. It is not visible for ion m/z 29 since it includes only two carbon atoms; the isotopic contribution of ^{13}C is in this case only about 2%.

9.6.2.2 Chlorine Isotopes

The chlorine element is naturally present as ^{35}Cl (76%) and ^{37}Cl (24%). Generally speaking, the number of peaks of the isotope pattern equals the number of chlorine atoms +1. For an RX_n ion containing n atoms of an element X, the relative abundances of the isotopic peaks are given by the formula $(a + b)^n$, where a is the abundance of the lightest isotope of X and b is the abundance of its heaviest isotope. The lightest chlorine isotope is ^{35}Cl. Its abundance is 76% or 0.76; the heaviest isotope is ^{37}Cl; its abundance is 24% or 0.24.

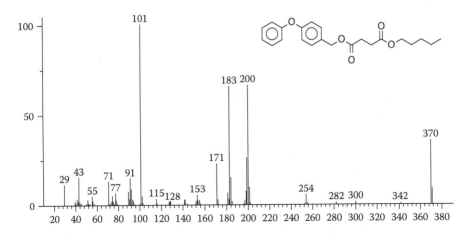

FIGURE 9.67   Electron ionization mass spectrum of pentyl 4-phenoxybenzyl succinate.

If an ion contains 1 chlorine atom, the isotope pattern presents $1 + 1 = 2$ peaks. The relative abundances of each of the peaks are calculated as $(a + b)^1 = a + b$:

1 peak at R + 35 of intensity 0.76, base peak of the isotope pattern
1 peak at R + 37 of intensity 0.24 corresponding to $0.24/0.76 = 31\%$ of the base peak

Figure 9.68 shows the EI mass spectrum of a monochlorinated chlorobenzoquinone molecule. The relative proportions of certain ion couples allow rapid identification of the chlorinated ions: m/z 142/144, 114/116, 88/90, and 60/62. On the same principle, we note that ion m/z 79 is not chlorinated due to the absence of a peak at m/z 81. The case of the m/z 86 ion is more complex. Careful observation of the isotope pattern m/z 88/90 shows that the abundance of ion m/z 88 is proportionally more than it should be in theory for a monochlorinated ion. It is therefore likely that some of the m/z 88 ions constitute the contribution of ^{37}Cl isotopomers of isotope pattern m/z 86/88.

If an ion contains 2 chlorine atoms, the isotope pattern presents $2 + 1 = 3$ peaks. The relative abundances of the peaks are given by the formula $(a + b)^2 = a^2 + 2ab + b^2$:

1 peak at R + 70 of intensity $0.76^2 = 0.58$, base peak of the isotope pattern
1 peak at R + 72 of intensity $2 \times 0.76 \times 0.24 = 0.36$ or 67% (0.36/0.58) of the base peak
1 peak at R + 74 of intensity $0.24^2 = 0.06$ or 11% (0.06/0.58) of the base peak.

Figure 9.69 shows the EI mass spectrum of a dichlorinated molecule: dichloroethene. The relative proportions of the ions in the isotope pattern m/z 96/98/100 instantly reveal the presence of three chlorine atoms. The observation of the isotope pattern m/z 61/63 shows that the latter corresponds to a monochlorinated ion.

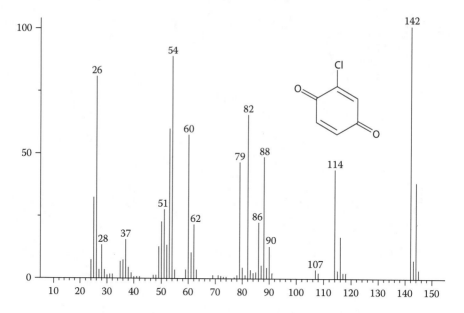

FIGURE 9.68 Electron ionization mass spectrum of chlorobenzoquinone.

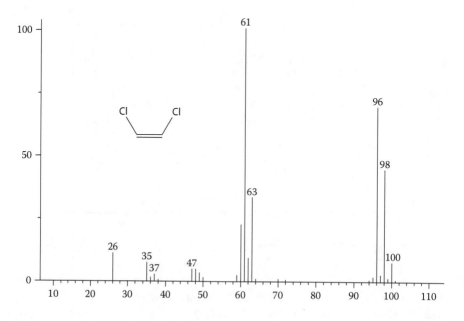

FIGURE 9.69 Electron ionization mass spectrum of dichloroethene.

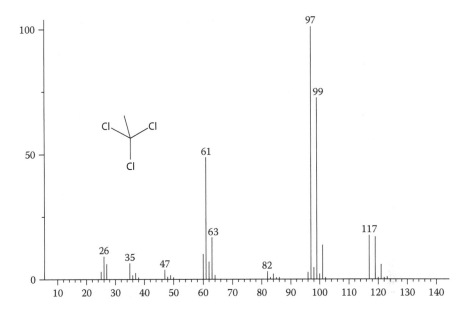

FIGURE 9.70   Electron ionization mass spectrum of trichloroethane.

If an ion contains 3 chlorine atoms, the isotope pattern presents $3 + 1 = 4$ peaks. The relative abundances of the peaks are given by the formula $(a + b)^3 = a^3 + 3a^2b + 3ab^2 + b^3$:

1 peak at R + 105 of intensity $0.76^3 = 0.44$, base peak of the isotope pattern
1 peak at R + 107 of intensity $3 \times 0.76^2 \times 0.24 = 0.42$ or 95% (0.36/0.44) of the base
   peak
1 peak at R + 109 of intensity $3 \times 0.76 \times 0.24^2 = 0.14$ or 32% (0.14/0.44) of the base
   peak
1 peak at R + 111 of intensity $0.24^3 = 0.014$ or 3.2% (0.014/0.44) of the base peak

Figure 9.70 shows the EI mass spectrum of a trichlorinated molecule: trichloroethane. We can distinguish an isotope pattern corresponding to three chlorine atoms: m/z 117/119/121/123 (the peak of the 123 ion is difficult to see because of its weak abundance); a pattern corresponding to 2 chlorine atoms: m/z 97/99/101; and a pattern corresponding to a monochlorinated ion: m/z 61/63.

9.6.2.3   Bromine Isotopes

Like chlorine, bromine is present naturally as two isotopes: ^{79}Br and ^{81}Br, in respective relative proportions of 51% and 49%. The principles described for chlorine also apply to bromine. Only the proportions of isotopomers in the isotope patterns differ. If an ion contains 1 bromine atom, the isotope pattern presents 2 peaks:

1 peak at R + 79 of intensity 0.51, base peak of the isotope pattern
1 peak at R + 81 of intensity 0.49 or 96% (0.49/0.51) of the peak at R + 79

If the ion includes 2 bromine atoms, the isotope pattern presents 3 peaks:

  - 1 peak at R + 158 of intensity $0.51^2 = 0.26$ or 52% (0.26/0.50) of the peak at R + 160
  - 1 peak at R + 160 of intensity $2 \times 0.51 \times 0.49 = 0.50$, base peak of the isotope pattern
  - 1 peak at R + 162 of intensity $0.49^2 = 0.24$ or 48% (0.24/0.50) of the peak at R + 160

If the ion contains 3 bromine atoms, the isotope pattern presents 4 peaks:

  - 1 peak at R + 237 of intensity $0.51^3 = 0.13$ or 34% (0.13/0.38) of the peak at R + 239
  - 1 peak at R + 239 of intensity $3 \times 0.51^2 \times 0.49 = 0.38$, base peak of the isotope pattern
  - 1 peak at R + 241 of intensity $3 \times 0.51 \times 0.49^2 = 0.37$ or 97% (0.37/0.38) of the peak at R + 239
  - 1 peak at R + 243 of intensity $0.49^3 = 0.12$ or 32% (0.12/0.38) of the peak at R + 239

Figure 9.71 displays the EI mass spectrum of a tribrominated molecule: tribromo-ethene. We can distinguish an isotope pattern corresponding to three bromine atoms: m/z 262/264/266/268 (M$^{+\bullet}$); two patterns corresponding to 2 bromine atoms: m/z 183/185/187 and m/z 158/160/162; and three patterns corresponding to monobro-minated ions: m/z 104/106, m/z 91/93, and m/z 79/81 (Br^{+}).

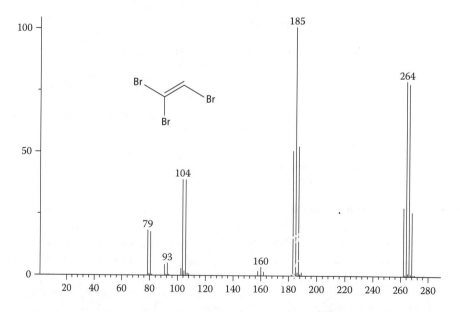

FIGURE 9.71   Electron ionization mass spectrum of tribromoethene.

A few years ago, the determination of the shape of an isotope pattern of a multi-halogenated ion was a classic university mass spectrometry exercise. Now, many free programs available online allow the instant determination of the shape of an isotope pattern of any ion from its raw formula. The earlier exercise is no longer of interest but a spectrometrist should understand the main principles. The calculations above remain applicable to ions carrying large numbers of halogens but the next section will demonstrate that not all isotopomers may be observed.

9.6.2.4 Incomplete Isotopic Distributions

The greater the number of isotopomers, the more peaks are included in an isotope pattern and the more the ionic signal is diluted. Beyond a certain number of iso-topomers, some ions become too few for detection. As an example, consider the isotopic pattern corresponding to the molecular ion of hexachlorobutadiene in Figure 9.72. With six chlorine atoms, the isotope pattern should present seven main peaks (m/z 258, 260, 262, 264, 266, 268, and 270) in addition to the peaks of ^{13}C. In reality, the m/z 270 ion does not appear in the mass spectrum. The reason is that the relative abundance of the isotopomers containing six ^{37}Cl atoms among the isotopomers constituting the molecular ion is 0.24^7 or 4.6×10^{-5}. This corresponds to a number of ions in the source below the limit of detection of the mass spectrometer.

9.6.3 Nitrogen Rule

The nitrogen rule is: if a compound contains an even number of nitrogen atoms (zero included), its molecular weight is even. Consequently, a molecular ion possesses an even m/z ratio. The ions issued from its fragmentation will have odd m/z ratios if

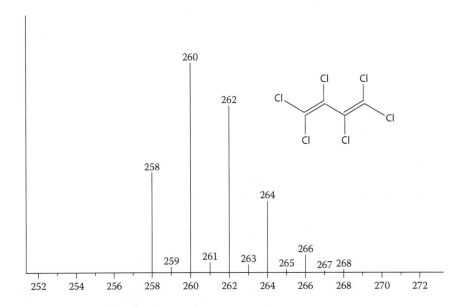

FIGURE 9.72 Isotopic pattern corresponding to molecular ion of hexachlorobutadiene.

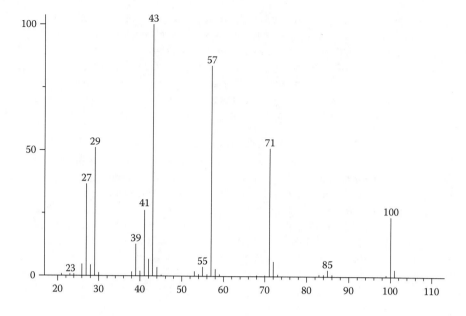

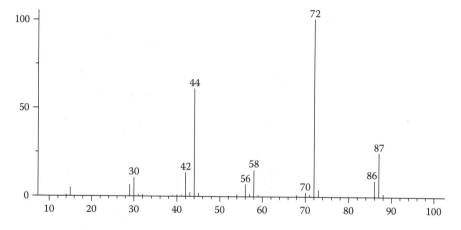

FIGURE 9.73　Mass spectra of 3-hexanone (top) and methyldiethylamine (bottom).

they result from simple (α and σ) cleavages and even m/z ratios if they result from rearrangements. The reasoning is obviously the same for molecules with odd numbers of nitrogen atoms that supply molecular ions with odd m/z ratios, ions issued from simple cleavages of even m/z ratios, and ions from rearrangements of odd m/z ratios. Figure 9.73 shows the spectra of 3-hexanone and methyldiethylamine.

Because 3-hexanone does not contain any nitrogen atoms, the molecular ion has an even mass/charge ratio: m/z 100. The fragment ions observed in its spectrum are m/z 27, 29, 41, 43, 57, and 71. Their odd m/z ratios indicate that they issued from simple cleavages (and subsequent fragmentations); none is the product of a rearrangement.

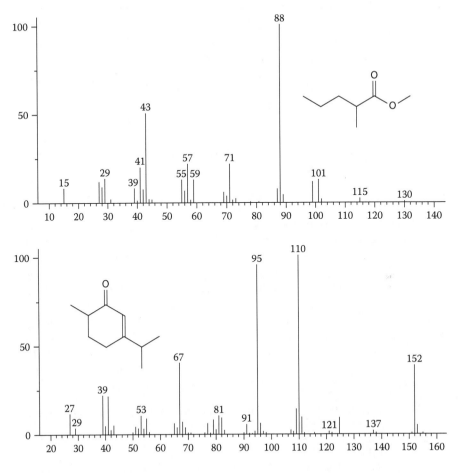

FIGURE 9.74 Mass spectra of methyl 2-methylpentanoate (top) and 3-isopropyl, 6-methyl, 2-cyclohexen-1-one (bottom).

Methyl diethylamine contains a nitrogen atom. Consequently, the molecular ion is of an odd mass/charge ratio: m/z 87. The fragment ions observed in its mass spectrum have even m/z ratios (30, 44, 58, 72, and 86), so we can conclude that none of them is the product of a rearrangement.

Figure 9.74 displays the mass spectra of methyl 2-methylpentanoate and 3-isopropyl,6-methyl,2-cyclohexen-1-one. Methyl 2-methylpentanoate does not contain any nitrogen atoms and thus the molecular ion has an even mass/charge ratio: m/z 130. This ion is absent from the mass spectrum, as is often the case with esters. Among the fragment ions observed in its spectrum, the m/z 88 ion of an odd m/z ratio results from a rearrangement. The only rearrangement allowed by the structure of the molecular ion is the McLafferty rearrangement. The 3-isopropyl,6-methyl,2-cyclohexen-1-one does not contain nitrogen atoms either; the molecular ion of even m/z 152 is observed on the spectrum. The fragment ion of even m/z 110

m/z 152   m/z 124

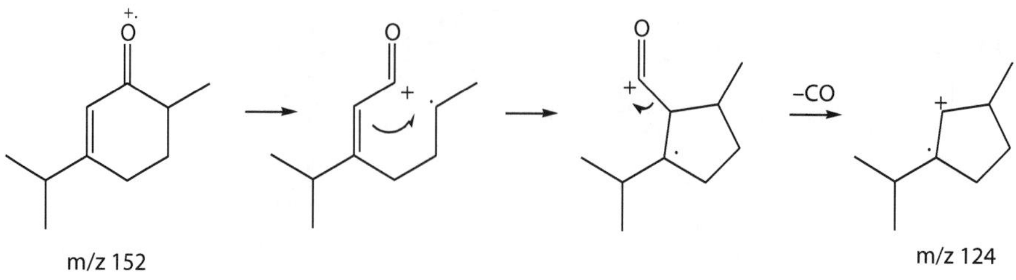

FIGURE 9.75 Formation of the m/z 124 ion from the molecular ion of 3-(isopropyl), 6-methyl, 2-cyclohexen-1-one.

therefore issued from a rearrangement. The structure of the molecular ion allows a retro Diels-Alder reaction.

The spectrum of 3-isopropyl,6-methyl,2-cyclohexen-1-one is interesting because it contains the trace of an even ion at m/z 124 that results from the elimination of carbon monoxide from the molecular ion. This loss of CO results from an α cleavage, followed by an electron transfer that leads to the stabilization of the radical (goes from secondary to tertiary) and the formation of a five-membered ring (see Figure 9.75). The spectrum shows the presence of an ion in a minority (of the same parity as the molecular ion) that did not result from a McLafferty rearrangement or a retro Diels-Alder reaction. Fortunately, the ions issued from these rearrangements are usually not abundant in mass spectra and this avoids ambiguities about their origins.

REFERENCES

1. Bouchonnet, S. 2012. *L'interprétation des spectres de masse en couplage GC-MS—cours et exercices corrigés.* Cachan: Lavoisier.
2. Haynes, W. M. 2012. *CRC Handbook of Chemistry and Physics,* 93rd ed. Boca Raton, FL: CRC Press.
3. Lias, S. G., J. E. Bartmess, J. F. Liebman et al. 1988. Gas-phase ion and neutral thermochemistry. *J. Phys. Chem. Ref. Data* 17 (Suppl. 1).
4. Lide, D. R. 2009. *CRC Handbook of Chemistry and Physics*, 90th ed. Boca Raton, FL: CRC Press.
5. NIST Chemistry WebBook. 2012. http://webbook.nist.gov/chemistry/.
6. Bouchonnet, S., S. Kinani, Y. Souissi et al. 2011. Investigation of the dissociation pathways of metalachlor, acetochlor and alachlor under electron ionization: application to the identification of ozonation products, *Rapid Comm. Mass Spectrom.*, 25: 93–103.
7. Petrucci, R. H., W. S. Harwood, and F. G. Herring. 2002. *General Chemistry,* 8th ed. Upper Saddle River, NJ: Prentice-Hall.
8. Harrison, A. G. 1992. *Chemical Ionization Mass Spectrometry.* Boca Raton, FL: CRC Press.
9. Dougherty, R. C. 1981. Negative chemical ionization mass spectrometry: applications in environmental analytical chemistry. *Biomed. Mass Spectrom.* 8: 283–292.
10. Kontsas, H. and K. Pekari. 2003. Determination of polychlorinated biphenyls in serum using gas chromatography–mass spectrometry with negative chemical ionization for exposure estimation. *J. Chromatogr. B* 791: 117–125.
11. Laramee, J. A., B. C. Arbogast, and M. L. Deinzer. 1986. Electron-capture negative ion chemical ionization mass spectrometry of 1,2,3,4-tetrachlorodibenzo-para-dioxin. *Anal. Chem.* 58: 2907–2912.

Index